Martha's Vineyard Decoys

MARTHA'S VINEYARD DECOYS

Stanley Murphy

Photographs by George Moffett

DAVID R. GODINE, PUBLISHER BOSTON

David R. Godine, Publisher
306 Dartmouth Street, Boston, Massachusetts 02116

LC 78-58592
ISBN 0-87923-260-9

Printed in the United States of America

Dedication

To George M. Moffett, Jr.
Blue-water sailor, naturalist, photographer,
gentle and generous public and private benefactor.

Preface

My interest in decoys began with a January walk in what was known then (1956) as Iva's Woods, several hundred yards from the dunes of the ocean beach at Quansoo, on the south shore of Martha's Vine-Yard. Years earlier a hurricane tide had pushed a thick line of flotsam far inland and a part of this line ran through the isolated woods. Half buried in the decaying refuse I found a grey, part-rotted duck decoy, its paint stripped away by the weathering of many seasons, and when I examined it I realized that it was carefully hand-carved by someone who knew a great deal about birds, someone who went beyond craft and created a work of art.

I took it home and cleaned and patched it and painted it for a bluebill drake. Later, beginning a serious study of decoys, I discovered that it had been made by Keyes Chadwick of Oak Bluffs.

All through that winter I spent the evenings and weekends seeking out more hand-carved ducks, so that by the arrival of springtime I had all of the decoys there was room for and as suddenly as the collecting obsession had struck, it faded away. While the acquisitive interest was keen it led me to some of our oldest wildfowlers and to some of their grand reminiscences, regrettably few of which I committed to written notes. Regrettably, because those men have since departed this world, every one of them, and with them has gone the last chance of identifying positively the work of certain carvers of their time and before.

In the intervening years, as collector interest has grown and as old decoys have begun to assume their rightful high place in the folk art of this country, and as a few good books dedicated to them have been published, I have become increasingly aware of the singular position of Martha's Vineyard in this field. There is no clear-cut Vineyard style in the sense that there is an easily recognizable Barnegat Bay, Delaware River, Chesapeake Bay or Illinois River style, to mention just a few. Certain particularly talented men influenced the work of a few others, to be sure, but most of the hunters of the great old days of Vineyard wildfowling were not imitative and were too hard up for cash to buy the factory decoys that may have been available. And the unusually rich concentration of ducks and geese on this relatively small island led many of the resident men to become gunners. This resulted in a body of decoy carvings of more than average interest and it is my aim now to present some of these birds to a wider audience and to identify the men who created them wherever possible.

Inevitably it will be brought to my attention that some decoy or decoys that I place in the "unknown carver" category will prove to be of off-Island manufacture. I apologize in advance for any such errors and welcome any clarification. Every decoy portrayed was found on Martha's Vineyard and has no known history that would indicate it originated elsewhere.

I hope that the near future will see the publication of other books dealing with the decoys of specific areas, done by authors most familiar with the territory. The excellent broad-spectrum works of Barber, Mackey, and Starr are a fine introduction to the world of decoys but the time has come to explore the subject more closely. The East coast alone, from Nova Scotia to Florida, can be broken down into many localities each deserving of a detailed study, each rich in decoy material still, in spite of the passage of time.

All of the photographs, except where specified, are the work of George M. Moffett, Jr. of Edgartown, for whose time, talent and patience I am deeply grateful. Most of them were taken in his house which overlooks an intimate view of Anthiers Pond and its marshes, a great tidal pond still known to some by the Indian name of Sengekontacket, where a number of decoys in this book were gunned over by men long since gone to a, presumably, happier hunting ground.

The decoys illustrated are the property of many different owners. I have thanked them personally for the loan of their carvings for photographing but that is poor compensation for their kindness and trust. It was decided that to name them here would be a dubious honor since it might attract the attention of collectors, dealers, or thieves. Contrary to the opinion of some, the three are not necessarily the same.

Contents

PREFACE vii
INTRODUCTION 3
THE KNOWN CARVERS 11
- Edward Francis Adams 13
- Henry Keyes Chadwick 27
- Max Chamberlain 59
- William E. Dugan 61
- Russell Hancock 71
- Robert L. Jackson, Hiram Crowell, and Manuel Swartz Roberts 75
- James Look 79
- Matthew H. Mayhew 83
- William H. Mayhew 85
- Allen Norton and Winthrop B. Norton 87
- Benjamin Warren Pease 91
- Frank Richardson 95
- Benjamin D. Smith 103
- Edwin T. Smith 111
- Alton C. Tuckerman 113
- Littleton C. Wimpenny 117

THE UNKNOWN CARVERS 119
- Shorebirds by Unknown Carvers 153

CONCLUSION 163

Martha's Vineyard Decoys

Introduction

I

Martha's Vineyard lies about four miles off the coast of Massachusetts, southwest of Cape Cod. It has the general shape of a triangle whose east-west base is twenty miles long, with a measure from top of triangle to base of nearly nine miles. It constitutes, along with the Elizabeth Islands to the northwest, the Commonwealth county of Dukes, and is divided into six townships: Edgartown, Oak Bluffs, Tisbury, West Tisbury, Chilmark, and Gay Head.

The eastern area of the Island is lowland and sandy, wooded chiefly with jack pine and oak. In the central plains are broad expanses of scrub oak surrounding a State forest of eastern white pine, while certain stretches of the north shore are heavily wooded with old stands of mixed hardwoods. The land rises gradually from the center westward, becoming hilly and boulder-strewn, with ridges formed by the alluvial deposits of the last glacier edges. The ground cover here is low—much of the acreage was once grazed by sheep—and miles of rugged stone walls still define old boundaries and fields. Knowing travelers are reminded of western Ireland and Scotland. The hill country culminates in the high, spectacular cliffs of Gay Head, where in storm times the sea is colored with ochres of clay that slowly erode from the banks.

An aerial view shows a long series of greater and lesser ponds some of which are periodically opened to the sea by man or by nature. The salt ponds, and a few that remain fresh, form a chain along the southern edge of the Island from Chappaquiddick to Gay Head and it was for use on these ponds, marshes, inlets and bays in the years from, roughly, 1870 to 1950 that the decoys in this book were carved.

The Vineyard was well qualified for the purposes of wildfowlers. The waters sustained all manner of food prized by the birds: seed clams in the salt ponds for the divers; succulent plants for the dipping ducks and geese in the fresh ponds; great beds of eel grass along certain shores for brant, and bait fish everywhere for the mergansers.

The salt meadows lured flocks of plover, while miles of tidal shore and marsh harbored the tiny organisms favored by such shorebirds as the yellowlegs, dowitcher and curlew. The Island was the jump-off point for the longest leg of the journey for migratory birds on the way south on the Atlantic flyway. Next stop, or landfall, would be Chesapeake Bay, or perhaps Albemarle, Currituck, or Pamlico Sound, and for some far wanderers such as the golden plover, South America. In the spring the Vineyard was one of the primary resting places on the flight north to the breeding grounds.

On the Canvasback:

> In my early experience the Canvasback was regarded as little more than a straggler in New England, though occasionally a few were taken. The number has been increasing, however, within recent years, and last year (1910) many were seen in the ponds on Martha's Vineyard . . .
>
> As it finds its favorite food in some of the ponds on Martha's Vineyard, this may account for the fact that it is more common there than elsewhere in Massachusetts.

On the Golden Plover:

> . . . the pastures of Tuckernuck, Nantucket and Martha's Vineyard were favorite feeding grounds of this bird.

In discussing extinct species, Labrador Duck:

> Audubon never saw the bird alive. The specimens from which his drawings of the species were made were shot by Daniel Webster at Martha's Vineyard, Mass., and are now in the collection of the National Museum at Washington, D.C.
>
> Freeman (1807) includes the Shoal Duck (Labrador Duck) as one of the species found on Martha's Vineyard.

On pursuing wildfowl in boats:

> . . . a reasonable amount of shooting from the shore will not disturb them (the ducks) if they are not pursued. It is largely due to a recognition of this fact, and to a special law prohibiting the pursuit of wild-fowl in boats, that Martha's Vineyard now has the best duck-shooting in Massachusetts. Formerly the gunners themselves observed an unwritten law forbidding the pursuit of fowl on the ponds. One man (an outsider), defying public sentiment, succeeded in driving most of the ducks out of one of the larger ponds in one day by pursuing them in a boat and shooting at them. A law resulted, prohibiting this pastime on the ponds of Martha's Vineyard.

So. If one wanted to shoot ducks in Massachusetts in 1910, and for some years after that date, Martha's Vineyard was a likely place to go. It was an area where many ducks were hunted and many decoys were needed to lure them to the hunter's gun.

In 1912 the ornithologist Edward Howe Forbush published his remarkable study, *Game Birds,*

*Wild Fowl and Shore Birds of Massachusetts.** In it he described the immense, almost unlimited numbers of waterfowl available to the gunner until shortly after the Civil War. It was then that the mass production of breech-loading shotguns and development of dependable rail transport from the towns near the marshes to the big city markets heralded the beginning of the end of the vast flights of earlier years.

Between 1880 and 1900 the records of some of the established gunning clubs in the Commonwealth showed a precipitous decline in the number of birds killed, the 1900 take being about ten percent of the total of the number recorded in 1880. And the decline was across the board, affecting every species. Still, it is interesting to note that in the year Forbush was gathering material for his book (1910), Martha's Vineyard was still blessed with a sizable population of waterfowl. Here are a few of his observations:

On the Redhead:

> . . . it is now found in considerable numbers in this state mainly on the ponds of Martha's Vineyard, where the wild celery and the redhead grass grow . . .
>
> Mr. Lewis L. Hill of Jamaica Plain says that Redheads and Bluebills are very plentiful at Edgartown, and that Bluebills have increased slightly "in the last five years." He believes that there are from five thousand to eight thousand ducks every year in Edgartown Great Pond. A party of four men got one hundred and ten Redheads and Bluebills in five hours, and many bags of twenty to fifty were made in the fall of 1908. In that year, he says, there were many more Bluebills than Redheads; in the last three or four years the reverse has been true. Mr. Henry V. Greenough of Brookline says that about 2500 Redheads and Bluebills come into the Edgartown and Tisbury great ponds in the fall from October 1 to 15; rarely more come and seldom many less. At daybreak every day they leave Edgartown Great Pond and fly to Tisbury Pond, where the "feed" is more to their liking, spend the day there and return toward night to Edgartown. Some stop over at Fresh Pond and Oyster Pond. The number has not decreased and about the same number of birds are killed each year. Mr. Charles H. Brown of Vineyard Haven stated before the legislative committee on Fisheries and Game, in 1910, that the ponds on the south side of Martha's Vineyard were broken open by the sea in 1815 and flooded with salt water, so that they remained salt for years. This changed the character of the vegetable growth in those ponds. Some of them remained salt longer than others which earlier became fresh or brackish. From 1872 to about 1878 Edgartown Great Pond was salt as a result of artificial opening. Redhead grass grows in Great Pond. Various pond grasses also grow in the ponds. More than twenty years ago Mr. Herman Strater introduced the so-called wild celery. About six or eight years ago it became plentiful there. Since that time the number of ducks on the Island has increased slowly, and the increase of Redheads has been particularly noticeable. Mr. Brown says that two thousand Redheads remained in Anthiers Pond during the coldest weather of January and February, 1910, and that he has seen more than ten thousand ducks in Edgartown Great Pond at one time, and perhaps two thousand in the other ponds in the same period.

*Issued by the Massachusetts State Board of Agriculture, 1912.

II

Of the hundreds of Vineyard-made decoys in collections spread across the nation, only the work of Keyes Chadwick of Oak Bluffs is well-known and recognizable to the average collector; mostly through information (and misinformation) in the few books about decoys published since Joel Barber's pioneer study of the subject. Chadwick had some peers on Martha's Vineyard though, and there was one man who surpassed him. There were several others almost in the Chadwick class, as well as a surprising number of men who turned out stylistically powerful carvings.

I discovered years ago that the older hunters had very little interest in the matter of who made what decoy. It didn't occur to them to be aware of such things because almost anyone with a little time on his hands could do the job; improvisation was a fact of life to the average Vineyard farmer-fisherman. Cash, in the golden years of decoy carving, was hard to come by, therefore even when a dozen factory blocks could be had for eight dollars, painted, a man would be apt to make his own. The materials could be found in driftwood on the beach or broken spars were available at the various boatyards—good cedar, well aged.

So the hunter would set out to carve a decoy on a winter's evening with the aim of attracting a live bird. It is a point to keep before us. Unlike the modern contest-carver, he wasn't attempting to impress his neighbor with his craft; he was out to con a living duck, or brant or goose or shorebird. In his mind's eye he knew what his prey looked like and in most cases had no need to consult a photograph or someone else's decoy. Occasionally, even often, the result was an excellent, unselfconscious work of art of primitive strength, usually capturing some salient feature of the desired bird.

The variety of Vineyard shelldrake decoys—the American and red-breasted mergansers—is a good illustration. One carver insisted on showing each tiny tooth in the bill, but no comb; another lovingly enhanced the comb but left the bill quite plain. One man thrust the head and neck forward in the alert way of a fast-swimming hen; his gunning partner had the head thrown back, the bill slightly raised, another characteristic movement of this species. But each man captured an essence. It is honest, deeply personal work whose special quality stems from the fact that just as each man saw differently, he carved the difference. It is true that a few of our best craftsmen were imitative of an admired predecessor, but with the majority of Island decoy makers it was every man for himself.

Artists today, in any medium, spend much of their time trying to pare away the veneer of photographic naturalism in a search for truly personal expression. For the old non-artists, the decoy carvers, there was no such search. Their originality was ready at hand, unblocked by artistic pretensions, and the frequent result of their efforts, unsought, was art.

The beauty and power of the old decoys is in their manifest differences of style. The greatness of any primitive (which is not to say unsophisticated) art forms—the bronzes of Benin, the totems of the Northwest coast Indian tribes, the masks of New Britain and Africa, to mention just a few—stems from the basic fact that they were created to function as utilitarian and religious objects. They were needed. The pale imitations of these same objects carved today by tribal descendants to gain the tourist or trade dollar fail as art. The strength and honesty, the mysterious truth of the original, is missing. So it is, for the most part, with the bulk of modern decoy carvers. The talent is there, abundantly there, but it is used merely to make something very pretty. And prettiness has nothing to do with art.

III

Anyone interested in decoys has noted the quantum jump in market prices in recent years, never mind the quality. It is pleasing to see that a more proper value is now being attached to some of these objects of folk-art. Certain of the decoys are priceless, and it is easy to foresee a time when the best examples of old, working decoys will be in museums, available to everyone. It is simply what happens to the finest works of art.

There is an increasing tendency among collectors today to rate the value of a given decoy according to the condition of its paint. Since I have few decoys from off-Island, I wasn't aware of this situation until quite recently, when I began to come in contact with mainland collectors who are interested in Vineyard birds.

The basic fact to consider, it seems to me, is that decoys are sculpture, and it should be mainly for its sculptural quality that a decoy is judged. The best paint job in the world won't make a poor piece of sculpture great, and the worst daubing cannot hide a really good carving.

I have never hesitated to remove entirely all paint from wretchedly overpainted or badly worn ducks. What is revealed is the original concept of the carver, the lovely grain of the old wood, perhaps with here and there a piece of #6 shot imbedded and maybe numerous cracks, splits, and nail-sick holes. And it is no less beautiful a sculpture than its sibling in pristine paint coat.

Certainly it is good to see a decoy in its original paint, and let me say here that a decoy overpainted several times *by its maker* is still in its original paint. Some of the carvers were wonderfully distinctive and able as painters too, but there is a clean, clear beauty in the stripped-down wooden bird that needs no apology in comparison with a fully painted example of the same bird by the same maker. There should be no vast difference in value if the decoy is intact.

Consider that the classic scupture of ancient Greece—the work of Praxiteles, Phidias, and their contemporaries—was fully and carefully painted after the marble was carved. No vestige of that paint has survived the centuries and it isn't missed. Even arms, legs, and an ear or two aren't missed. The sculptural greatness is there. The one example in my collection of the superb work of Elmer Crowell of East Harwich is an early black duck that was solidly covered with grass-green enamel, having served as a garden ornament in its last incarnation before I found it. It sits on my mantel now, stripped to the rich surface of its warm cedar and pine, enhanced by the respectable scars of all its years.

The collector should know by this late date that most antique dealers are inclined, with an exuberant disregard for accuracy, to attribute any and all decoys to Crowell or Wheeler or Chadwick or Hudson or to some known carver. It wouldn't be surprising if some were by now attributed to Rembrandt. In a way it is difficult to blame the dealers since source books to aid positive identification are sadly lacking, and a man likes to take pride in his stock in trade.

Therefore, if a prospective buyer's own knowledge of carvers is wanting, it would seem logical to me that he make a purchase only if the bird pleases him on its own merits, regardless of who may have carved it. There are many decoys of surpassing beauty whose makers will never be identified. There were relatively few men who carved professionally—who made a large body of birds for sale, or who needed a lot of decoys to stock a gunning club or a guide's rig. Most gunners made a few dozen or less for their own use only, and with the attrition of the years—breakage, loss, rot, fire, etc.—these men may be represented in this world by no more than two or three surviving decoys.

My respect goes to the collectors who have the courage of their own convictions. If you have to be *told* a thing is beautiful, desirable, interesting or whatever, forget it. One of the most admirable decoys I ever saw was being used as a doorstop by the wife of an old Island gunner. She wouldn't part with it for anything because she loved it and wanted it to remain right there, "next the door" as she put it. It was part of the fabric of her life. The man who has amassed two hundred "name" decoys at considerable expense, remaining blissfully blind to any esthetic quality in them, is something else again. But that, too, is often the way it happens with art.

IV

The commonest decoys to be found on the Island are the factory ducks, mostly Mason and Mason imitations, of the lower grades. It is a fairly safe assumption that the hand-carved birds which are the concern of this book were made either because the maker couldn't afford the store-bought article or because he thought he could make a better decoy, and sometimes a combination of the two. Only two men here are known to have made quantities of duck decoys to sell: Keyes Chadwick of Oak Bluffs, and Little-ton Wimpenny of Edgartown. One man, John L. Mayhew of Oak Bluffs, made shore birds for sale.

The greatest number of Vineyard-made duck decoys were bluebills, with redheads a close second. In a good redhead year many bluebills would be overpainted. A few years later they might become blue-bills again. Black duck decoys were common too, and this wariest of all ducks was pursued mostly after sundown. One ancient gunner pointed out to me some dark brown, cork-stuffed bags in a basket on his porch and said, "There's the best black duck toller of all!" In the evening, anchored with the bottoms up, the bags have the look and movement of feeding blacks. The men who used them had no concern for the niceties of the sport. The idea was to put food on the table; the more ducks one could clobber with one shot on the water, the better.

A number of whistler decoys were made here, and brant too. The brant date mostly from the years 1902-1910 and 1920-1930 when there were particularly lush banks of eel grass for this specialized feeder in the bay between Oak Bluffs and Edgartown. A certain amount, possibly no more than a sprinkling, of bufflehead, teal, canvasback, mallard, and ruddy duck decoys were carved and, finally, a surprising quantity of mergansers. (Not including the hooded variety. These handsome little ducks are not uncommon here, even today, but I have yet to see a locally-made lure in their image.)

The fact that few Canada geese were made on the Vineyard may seem surprising, inasmuch as the goose is today a common bird on the Island, but most of the goose gunning in the old days was done from camps especially set up for the purpose. Live decoys were used, and the art of flying the goslings—the hatch of the previous summer—while keeping the parent birds penned, was developed to a high degree of perfection here. Goose camps rimmed the entire southern shore of the Island from Pocha Pond on Chappaquiddick to Squibnocket Pond in Gay Head.

The wild birds, heading south over the Atlantic on their migratory route, were tempted into the great ponds by these raucous Judas's who would then return to the area of the blind where grain had been sprinkled on the beach and in the shallows. When the wild geese decided to fly or swim in to join the feast, a gang of hunters concealed in a long, camouflaged trench, would fire on signal, and slaughter—there is no other word for it—as many of them as possible.

In some cases slat, or barrel, geese were used. They were gigantic things, four times the size of a living goose, moored far out in the water to catch the wild traveler's eye and hopefully bring him close enough to see the tethered geese contentedly feeding in front of the waiting skirmish line.

The federal government outlawed the use of live decoys at a rather late date—1935—and with the end of live decoying, most wildfowlers had to be content with whatever geese happened by during a duck hunt. Few could afford the store-bought goose decoys and not many chose to try their hands at the home-made variety. Riparian owners might indulge in wooden geese since in those days they could be left in the blind without much fear of theft, but they were simply too much weight for the average gunner to carry through the pre-dawn marsh. It was hard enough to manage duck decoys, gun, shells, and all of the other paraphernalia of a wildfowler.

The established gunning clubs were able to stock wooden geese, though, and the evidence is that at least two large rigs here were made by noted off-Island carvers: Joseph Lincoln of Duxbury, Mass., and (curiously, considering the distance) Ira Hudson of Chincoteague, Virginia.

Many of the pen-raised live decoy geese were released to shift for themselves after their employment as shills was outlawed. They stayed here instead of taking up the ancient seasonal migration, and they have thrived. They are much in evidence in the up-Island skies and seem almost tame until the first guns of opening day. After that, the hunter needs all of the skills and subterfuges and endurance of his ancestors to bring one to table.

The Known Carvers

1. FRANK ADAMS. Adams carving one of his small paperweights in his West Tisbury shop. Note the patterns hanging above him and the 3-masted weathervane on the floor behind him.

Edward Francis Adams 1871–1944

"E. Frank Adams. Ship Weather Vanes, Duck Paperweights and Door Stops." Such was the text of an advertisement in the Vineyard Gazette, May 5, 1944. Elsewhere in the same issue is the report of Adam's death.

He was born in West Tisbury and, the Gazette reports, ". . . As a boy of 19 was apprenticed to a Boston firm to learn the machinists trade, and he took to the work naturally. He was adept at handling tools and had a native feeling for fine measurements and finish. But a boy who has been brought up on the shores of Tisbury Great Pond, following the fish and wild fowl with boat and gun, does not thrive in the confinement of a city shop and after a year had passed without any abatement of his homesickness, Mr. Adams returned to the Island."

Following fifteen years employment as a carpenter in the larger Vineyard towns he began the work of more than a decade as a wheelwright in the shop of his neighbor, Brandon Mayhew. Finally he opened a shop of his own behind his home on Music Street in West Tisbury and there he spent the rest of his working life building furniture and carving duck and goose paperweights and doorstops and, his obituary says, ". . . ship weathervanes which now swing in the breezes all over the world."

The Adams decoy ducks have smooth, simplified, cedar bodies, both hollow and solid, and the heads are finely carved pine. His widgeon (Plate 4) has the most accurately widgeon-like head I've seen on any old decoy of this species. The paint pattern, when intact, is usually dark with a blurred feather pattern painted wet into wet. On the other hand his fine working shore-bird decoys are often painted in minute detail. The most common of his creations still to be found on the Island are the paperweight ducks carved in all sizes from tiny miniature to life-size. They are usually painted in soft, arbitrary color patterns and I include some here (pages 44, 45) to indicate the variety of his talent. Adams' day-book from the last year of his life, a record of daily sales and orders received, still exists. A small gleaning follows:

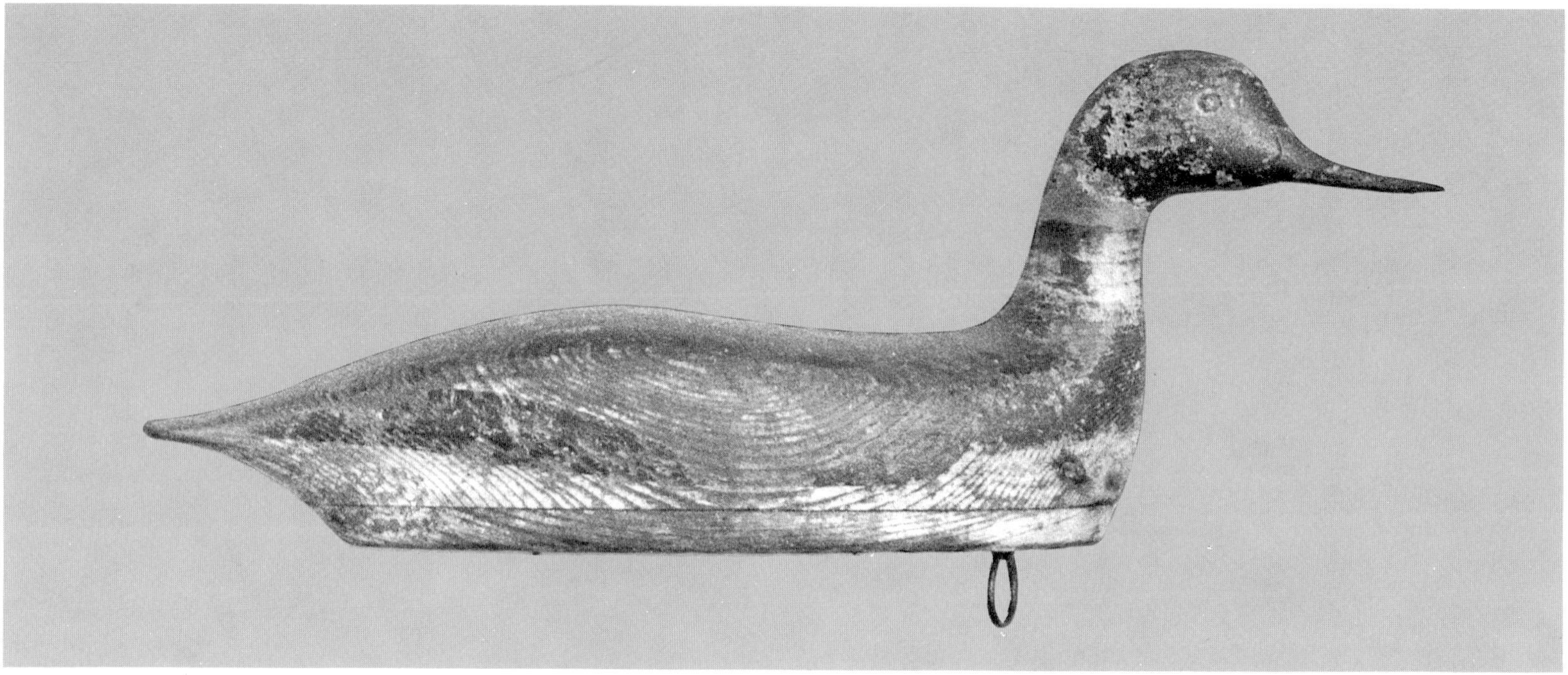

2. FRANK ADAMS. A red-breasted merganser male, a small, hollow decoy with head turned slightly to the right. The weight is a little rectangle of lead, countersunk and screwed to the bottom.

Tues. May 19 [1943]. Cloudy and foggy all day. Worked in shop painting spars and putting shrouds on two 3-masters.
Tues. Aug. 17. Nothing doing at all. Dry as a burnt boot.
Thurs. Aug. 19. Made six door-stops. Fine day.
Sat. Aug. 21. One duck and goose—$3.00.
Mon. Aug. 23. One duck 10 A.M.—$1.00.
One 3-master and cradle, about noon—$14.50.
Wed. Aug. 25. Two ducks in forenoon to folks from Edgartown on bicycles.—$2.75.
Thurs. Aug. 26. Three sets of callers. Never sold a *cents worth*.

One of the last entries is more to the purposes of this book: On Friday, August 24, Adams notes that he sold "three old decoys and one buoy—$3.00."

There is no indication that Adams made working duck or shorebird decoys for others. On the other hand it seems that he may have made some shorebirds which, though following his working pattern, were given or sold as a purely ornamental product in the years following the federal ban against decoying the little birds. A number of Adams shore birds exist in pristine condition.

3. *FRANK ADAMS. An early, hollow bluebill thickly overpainted a uniformly dull brown.*

We find no stylistic reverberation of Keyes Chadwick in the work of Frank Adams but it is a fact that the men were married to sisters, the Pease girls of West Tisbury, and that they visited frequently. When he began to carve professionally there would seem to be every reason for Adams to be at least somewhat imitative of his accomplished brother-in-law, but he managed to avoid it and remained entirely his own man, the creator of a style.

4. FRANK ADAMS. An excellent likeness of the baldpate, or American widgeon. The local hunters call them "whitebellies". and they are a fast, elusive target. Adams did most of his gunning on one of the few Vineyard ponds where they are regular visitors.

5. FRANK ADAMS. The baldpate of Plate 4. This decoy is hollow, has been repaired at one spot with a wooden plug, and has loose shot rattling inside.

6. FRANK ADAMS. Adams caught the essence of this hen very nicely. The method of cutting the eye into the wood and rounding the surface of it can best be seen here. When one could carve it that well, why use glass eyes?

7. FRANK ADAMS. Three late-period American Goldeneyes, "Whistlers". The drake is cut from a completely different pattern than the hens. The paint is original, bodies are solid.

8. FRANK ADAMS. The trio of whistlers from Plate 7. Back in their element after a long rest.

9. FRANK ADAMS. Three black-bellied plover in dark plumage for the Spring shooting. The bills of all are original, as is the paint. The eyes are the metal studs Adams used in his earliest decoys.

10. FRANK ADAMS. Another view of the plover in Plate 9. The bodies are much slimmer, side to side, than most of Adams' shorebirds.

11. FRANK ADAMS. Three graceful lesser yellowlegs, competently painted.

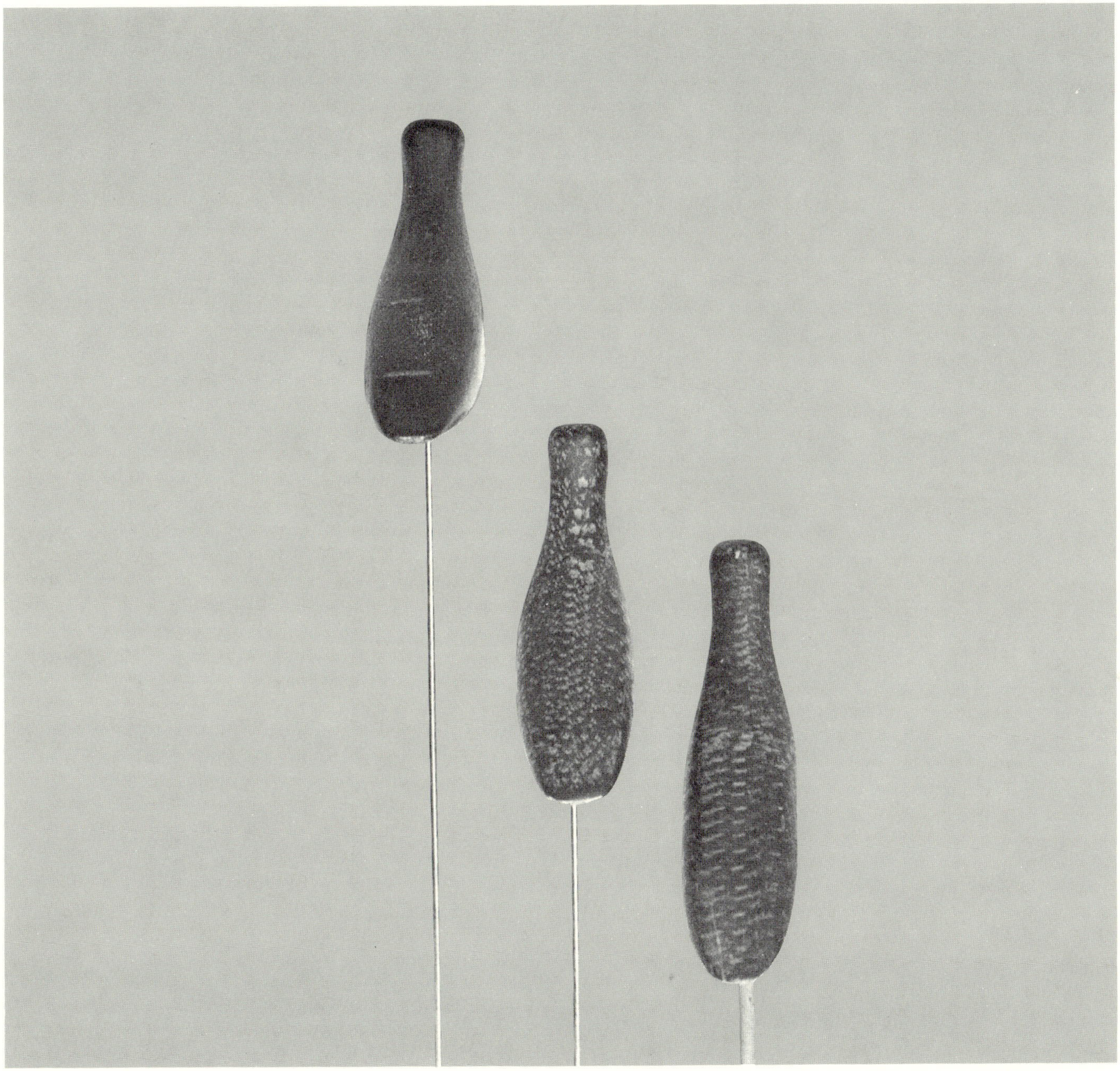

12. FRANK ADAMS. Another view of the yellowlegs in Plate 11.

13. FRANK ADAMS. A golden plover in transitional plumage. There was great variety in the flocks over the very long gunning seasons then permitted.

14. FRANK ADAMS. A fine shorebird, but precisely what kind of shorebird is the question. When two old gunners who shot many shorebirds early in this century were shown this carving they were uncertain as to what species it might be. But "... that would pull 'em in; most anything would. It was calling [imitating the whistle of whatever species appeared] that kept 'em comin'."

15. H. KEYES CHADWICK. c. 1950 at his Oak Bluffs, Massachusetts home.

Henry Keyes Chadwick 1865–1958

Keyes Chadwick was the only professional duck decoy carver of his generation on Martha's Vineyard. His life was long and productive, and a far greater number of his fine works survive than those of any of his contemporaries or predecessors.

He was born in Edgartown a few weeks before Grant and Lee met at Appomattox Courthouse. The family moved to Oak Bluffs when he was three years old and his education began in that town, and was continued at the New York State Agricultural College. He hoped to pursue a career of raising fancy poultry but instead was forced by economic necessity to become a carpenter in his home town. He wasn't able to devote much time to chicken-raising until around 1904 when he was hired for that purpose by a man named Owen who had a farm in Lambert's Cove, Tisbury.

Chadwick made his first decoys for his own use in 1881 when he was sixteen years old, then made no more for some years following. It was when he was temporarily disabled by an accident to his leg, and was unable to carpenter or hire out, that he began to concentrate more of his time on the production of decoys for sale. Even so, as he told a reporter many years later, he worked at carving only intermittantly, never steadily.

No readily identifiable ducks have come to light from his first attempts and it is only after he came under the influence of his fellow-carpenter and neighbor, Benjamin D. Smith, that we can begin to trace his development as a carver. Smith was an avid gunner who made decoys for his own use only and Chadwick recognized his mastery and adopted his patterns. The earliest known black duck carved by Chadwick (Plate 16) was copied from a Smith creation in its entirety, and he was then closely imitative of Smith in the working of heads of all species, but where Ben Smith appears to have experimented with flat-bottomed, solid and hollow ducks, Chadwick is not known to have carved a hollow decoy or one with other than a perfectly flat base.

16. KEYES CHADWICK. At bottom is the black duck of Plate 87 by Benjamin D. Smith. Above it is the early copy made by Chadwick, 1890-1900.

Prior to 1900 his decoys do not carry the distinctive flush-poured round lead weight; after that time it is always present. And the oldest Chadwicks are in some ways the most pleasing to the eye, with heads carved in full, careful detail (except for the nostrils, which seem always to have been painted on) (Plate 17). The necks tend to be thinner than in later work and the tails are deeply undercut and attenuated. As years passed the more delicate carving was eliminated. The tails, easily broken, gradually became more rounded and strong; necks, especially in the diving ducks, thickened; bill carving was simplified (Plate 18).

Chadwick made his last decoy in 1951, rounding out a creative span of seventy years during which from 1500 to 2000 wooden birds came from his hand. "Probably over 1500" was his own hesitant estimate; it seems likely to me that over 2000 is closer to the mark. In the course of the research for this

17. KEYES CHADWICK. A redhead drake, an early example. The eyes are painted and the area between the shoulders has been deeply carved. 1900-1920.

book I have seen over five hundred examples of his work, and those I have seen have in some cases been only the remnants of much larger Chadwick rigs. There is documented proof that even relatively late in his life he was selling in such quantity as 300, 191, and 100 to a lot, and these were ready-to-use decoys which were then put to work by hunters.

I am sure that there are many Chadwicks on the Island that I have not seen, and one of the few off-Island collectors I know told me that at one time he had as many as a hundred of them. Consider the number lost over the years to fire, breakage, and storms, and further consider that during the 1940's the Vineyard towns were visited by antique dealers who filled their pick-up trucks with heaping piles of used decoys (for which they paid as little as 25 cents each). A figure of over 2000 for Chadwick's total production is quite reasonable.

The changes from first to last are not dramatic. They came about through a slow process of refinement and resulted, finally, in rugged decoys that could take the hard usage of a gunning trip, that could be jostled one against the other in a gunny sack over a hunter's back without breakage. But the essential Chadwick "look" is in all of his decoys of every period. They are smoothly finished, graceful, and sometimes even stately birds.

When I first became interested in decoys, Keyes Chadwick was still alive and had been in a Vineyard nursing home for a year. I visited him there to talk to him of decoys and gunning, and during the two years remaining to him returned a number of times, usually carrying some wooden birds for him to identify or comment on.

18. KEYES CHADWICK. A pair of greater scaup (bluebills) in original paint. Chadwick's most common, most rugged decoys.

19. These Nineteenth Century wooden patterns were found in Chadwick's pattern box. All by the same hand, one is marked "Geo. F. Wing".

At the time this treasure was given to me I was interested in them only as a curiosity. There was no idea of ever trying to write about Keyes Chadwick or any other carver, and while the actual decoys fascinated me, I found little esthetic reward in the patterns. Several years went by before I talked one day with another of the fine old craftsmen of this Island, Manuel Swartz Roberts, in his shop in Edgartown. He had retired from the business of boatbuilding and spoke of wanting to make some decorative bird carvings to sell. I asked him if he would like to have the Chadwick patterns and he was delighted at the prospect, so after culling out a few items I wanted to keep, I took the boxes to him.

Within a year, he too went to his last reward. And a year beyond that, when I stopped at his house to inquire of his widow if by chance she knew the whereabouts of the patterns, I was told they'd been taken to the town dump by her son and thrown into the fire pit. *Sic transit gloria mundi.* How easily they would solve certain problems of Chadwick identification today.

Recently, to my astonishment, I discovered that a decoy collector on the Cape owns a third box of patterns (Plates 20, 21, 22). I believe them to be the ones Chadwick used in the latter part of his life as well as some chosen old favorites. The sleeper black duck (Plates 23 and 24) is there (the decoys made from it all date from 1950, give or take a year) and the canvasback head from which he made paperweights late in his career.

MATERIALS AND TECHNIQUES

The decoy bodies were cut from white cedar posts which had been stacked outside of the shop to cure for at least a year before use. Heads were made of sugar pine, a wood that takes and holds detail carving easily and yet is tough enough to withstand hard usage. To fasten the head to the body, Chadwick bored a hole, generally ¾ inch diameter, up from the bottom to within ¾ inch of the flattened area which would receive the head. A hole the diameter of the three-inch brass screw was drilled through the remaining cedar and, when the head was securely screwed onto the shoulder-platform, the hole in the bottom of the decoy was corked flush with the surface.

A hole was bored to receive the lead weight at the proper balance point in the long axis of the base, its diameter varying according to the size of the decoy. A large-headed nail was driven part way into the center of the hole and this served to secure the lead which was melted on the wood-stove in the Chadwick kitchen and poured until it was flush with the bottom. As neat, solid, and unobtrusive a method of weighting a decoy as has been invented.

His habit was to carve a series of bodies and follow them with assorted heads—it wasn't a matter of completing one bird before going on to the next.

The oldest Chadwicks have eyes that are metal studs. Some were simply painted on. Later, and for a considerable period, glass eyes are set in sockets which are purposely countersunk so as to leave a depressed rim about 1/16 inch in width around the eye. In the decoys of the latter part of his life, eyes are set at the precise level of a live duck's eye, in a hole the exact diameter of the eye.

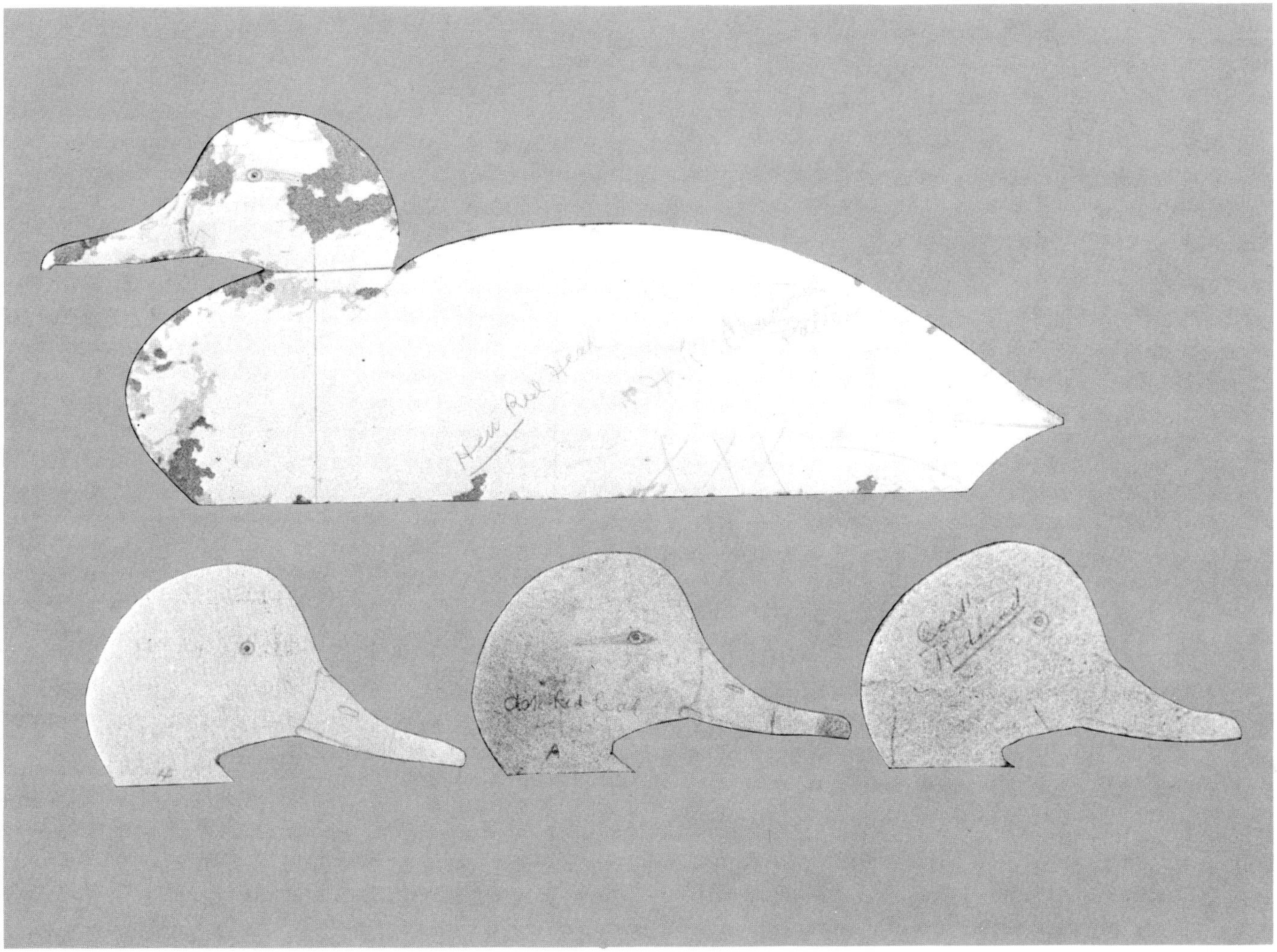

20. KEYES CHADWICK. Just a few of the many redhead pattern variations. These are cut from thin cardboard. Penciled notes on the body pattern suggest that thirty decoys of this type were made. I have seen perhaps six of them.

21. KEYES CHADWICK. A brant, five widgeon, and two redheads (lower left and right).

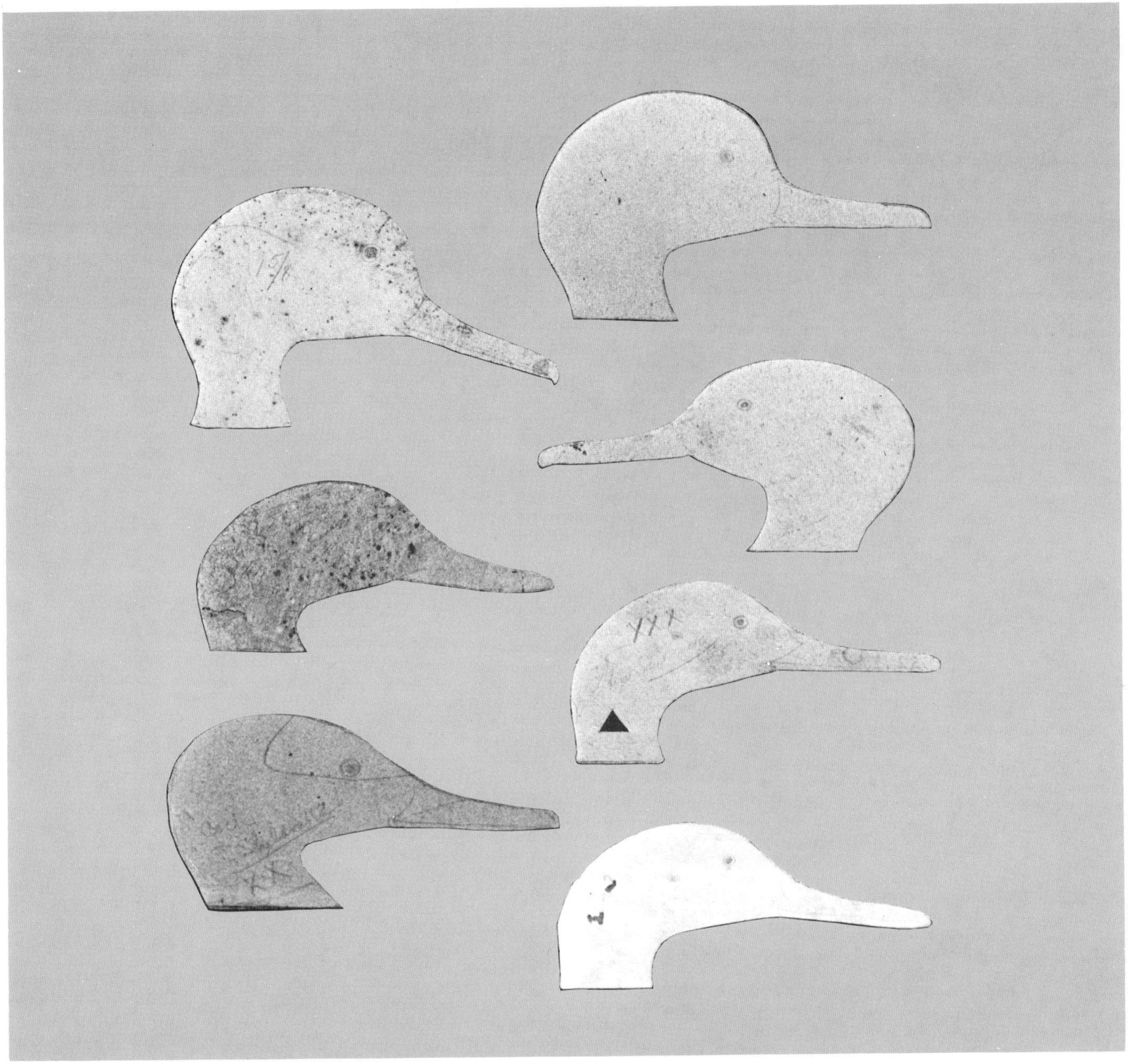

22. KEYES CHADWICK. Cardboard merganser patterns that cover a span of many years of carving. Every one was used, in some cases dozens of times.

23. KEYES CHADWICK. A black duck "sleeper". This one was painted by Chadwick and painted over by its owner. Perhaps a dozen more sleepers were made from the same pattern, but they were unpainted and unweighted and never got beyond the mantelpiece.

In his last year, Chadwick had his caretaker deliver to me, as an unannounced gift, two boxes full of his older patterns. I was surprised and delighted, and amazed by the number—literally hundreds—of variations. It was apparent that he was never totally satisfied with a given pattern, that his search was for perfection, a perfection always liable to further perfection. Bodies were lengthened or shortened or narrowed; the bottom area was changed; heads were raised or lowered. The variance between one redhead body and another might be in a quarter-inch drop in the small area just behind the neck but the changes were carefully mapped out on cardboard or paper or shingle stock first. Head patterns were varied in almost infinitesimal ways as well as by major alterations. There was much variety in the merganser heads, for instance, and these shelldrake decoys are among the loveliest of Chadwick's creations.

In one of the boxes there were two carefully preserved real duck heads—a bluebill and a redhead. And Chadwick caught their essential differences in his carving: the slight peak of the redhead's crown—not often observed by other carvers—as well as the obvious difference in the bills.

Also in this box was a group of old, thin wooden patterns by another hand—shelldrakes, redheads or bluebills, and a black duck (Plate 19). Obviously they are pre-Chadwick, but whose? The divers and

24. KEYES CHADWICK. Close-up of the "sleeper".

black duck are very much like early Chadwick heads, but the shelldrakes are wholly different. Primitive, and still very fine. Carefully penciled in print on the redhead is the name, GEO. F. WING. It is not in Chadwick's familiar hand.

Some of the paper and cardboard patterns were never used, but most of them show the black of a pencil tracing completely around the edge, clear proof that they resulted in at least one decoy duck, probably more.

He used the best available glass eyes and they were the proper color of the species involved. At the very end, to judge from the last sizable batch of decoys he made (which were purchased unpainted and unweighted by an Edgartown man after Chadwick ceased all production), his failing vision betrayed him into setting red eyes in bluebill heads, and eyes of different color in the same head. Except for this aberration, though, the last birds, particularly the marvelous widgeon (Plate 25), are of his usual high standard.

25. KEYES CHADWICK. A widgeon (baldpate) drake. One of Chadwick's last carvings, dated and initialed by him on the bottom: "H.K.C. 1950". He was 85 at the time, and didn't get around to painting or weighting it.

Color Plate A. BENJAMIN D. SMITH. Another Red-Breasted Merganser in original paint but a very different conception from Smith's earlier birds. This solid, round-bottomed beauty carries his unique weight. The eyes are glass. Both the tail and the bill were repaired by an expert hand after they were chewed by a puppy.

Color Plate B. BENJAMIN D. SMITH. *The head of a Red-Breasted Merganser hen in original paint. To me, one of the all-time great duck carvings.*

Color Plate C. BENJAMIN D. SMITH. A Red-Breasted Merganser drake in original paint. An early example of this master's work, with subtle carving between the shoulders, flat bottom, and painted eyes.

Color Plate D. KEYES CHADWICK. *An American Merganser hen, a rarity among old bird decoys. The paint is original.*

Color Plate E. KEYES CHADWICK. A Red-Breasted Merganser drake from Chadwick's great middle period. One of the "snake-head" variety in original paint.

Color Plate F. FRANK ADAMS. *A complete stool of Golden Plover in pristine original paint, and on their original stakes. The background is Tisbury Great Pond, where Adams leased property and did much of his gunning. This set was purchased from Adams by a man who went off to World War I before he could use them. By the time he came home, the 1918 law forbidding the shooting of shorebirds was in effect.*

Color Plate G. FRANK ADAMS. *Paperweights and doorstops, which Adams made in all sizes, included here because of their popularity with collectors and because they show a taste for fantasy, especially in the painting, that Adams never indulged in his decoy work.*

Color Plate H. KEYES CHADWICK. Redhead drake. A fine, early example of this species, carved when the influence of Ben Smith was still strongly evident in Chadwick's work. Original paint.

26. KEYES CHADWICK. Red-breasted merganser drake in original paint. 1900-1920. Chadwick had a separate pattern for the white wing area to guide his painting.

A man who watched Chadwick carve many decoys reported to me that he shaped the bodies, down to the final smoothing, with a hand axe. I asked if he were certain it was an axe and not a drawknife: "Yes sir!" he replied. "I tried a few strokes on a fresh block in his shop once, and cut too deep and spoiled the block. Keyes gave me hell." I will add that I saw keen and ready drawknives hanging in his shop. They had seen a lot of use. When Chadwick was interviewed in his eighty-fifth year he told the reporter that he roughed out the decoys with an axe, then used various knives and finally sandpaper. No machine tools except for jigging the basic outline drawn on the block from a pattern. A Chadwick decoy is a very smooth affair from which tool marks are absent.

He had a low opinion of his ability to paint his decoys, but the fact is that he was adequate enough to the task. Rather than blend one edge into another, he left the clear division line, and realistic feathering remained a mystery to him. He carefully cut paint patterns in thin cardboard for tricky areas such as the drake red-breasted merganser wing (Plate 26), the wings of the cock whistler, and the back designs of other ducks.

A certain amount of misinformation about the connection between Keyes Chadwick and Elmer Crowell of East Harwich has been published, mostly to the effect that Crowell did Chadwick's painting for him. The facts are these: That Chadwick once sent several black ducks over to the Cape to be painted by Crowell: that when they came back he pronounced them the most beautifully painted decoys he'd ever seen; that later he sent a few widgeon to be painted—no more than six, possibly only four—and that they were the sum total of Chadwick-owned decoys painted by the master, Crowell (Plate 28).

What is not generally known is that Elmer Crowell made a number of trips to Oak Bluffs for the purpose of buying Chadwick's work, unpainted. He would look over the latest batch and pick several dozen

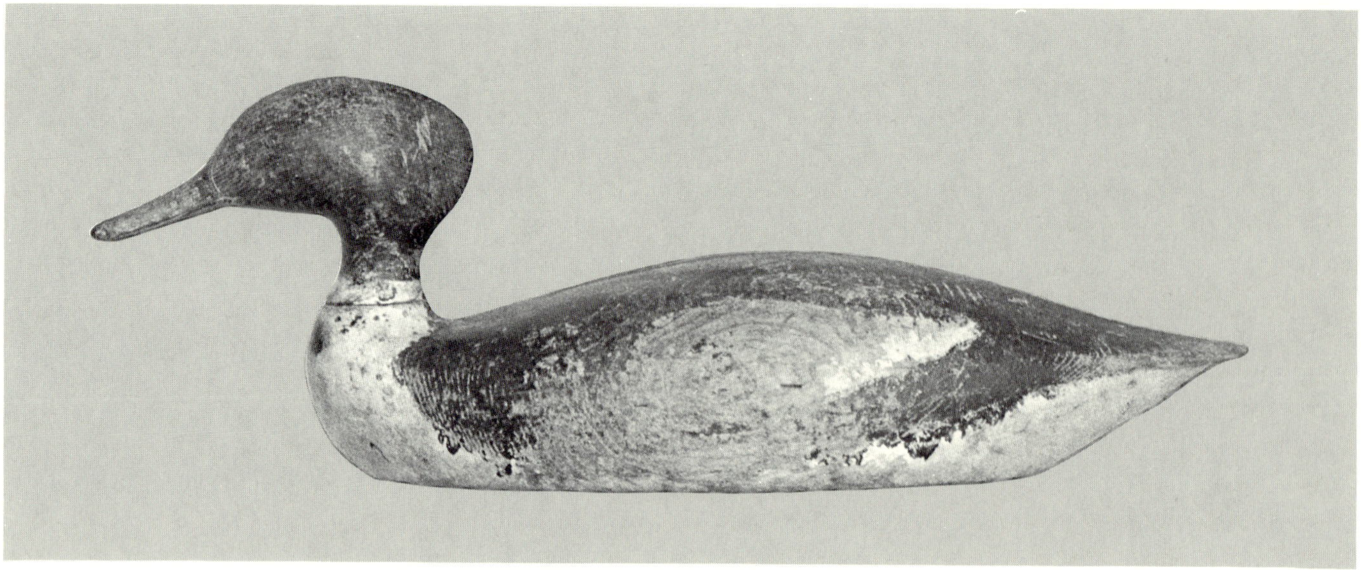

27. KEYES CHADWICK. An early red-breasted merganser drake with very worn original paint. Painted eyes. 1900-1910.

of the best, saying "I want that one, and that one, etc.," and Chadwick, not about to accept this careful pruning, would reply "All right, you can have them but you'll have to take a dozen of these and a dozen of those too."

The Crowell purchases were then taken to the Cape where he would paint the ducks in his inimitable fashion, and sell them from his shop. There was, of course, no intent at deception. Crowell was doing a large volume business at that time and couldn't supply the demand, hence the Chadwick decoys, and more stock in hand. And more customers satisfied.

It should be known that Crowell had the highest regard for his Vineyard peer. He once said to Chadwick, with witnesses present, "Chad, you are a better carver than I am," and then added: "but you can't paint worth a damn!"

I wonder now whatever happened to all of those Crowell-painted Chadwicks. Could it be that Crowell might have embellished or trimmed or re-cut parts of the Chadwick blocks to conform to his own style? Or did he sell them, after painting, without adding his famous brand? I have yet to see a branded Crowell that is anything but wholly Crowell (in the head carving at any rate). But I am not well enough acquainted with his work to rule out the possibility that some are altered Chadwicks.

Chadwick made his wooden birds in the basement shop of his home and, when the weather was moderate enough to allow it, he worked under the trees in the back yard. I was permitted by his caretaker to see the shop in that last year of his life. His tools were ready for use, clean, with wood handles rubbed to

28. KEYES CHADWICK. A pair of Chadwick widgeon painted by Elmer Crowell. The white dots are spatterings of ceiling paint of a later date.

a burnished smoothness by the varying grips of many years. The knives were razor-edged. The sweet smell of countless sweepings of cedar shavings lingered still.

He told me that there were three things he considered his special accomplishments: he raised prize-winning fancy fowl; he did fine Spencerian pen work of the sort used on diplomas and important documents; he made a good duck decoy. Of the three he was proudest by far of his record as a producer of fancy fowl. In an interview for the Vineyard Gazette in 1950 he said "I've never done what I wanted to do, and that was to raise chickens, just to raise lots of champion chickens." In 1953 he told another reporter that ". . . he did not want to be quoted as being a decoy maker. He was not, and that was all there was to it. He was a chicken raiser, and did that because he liked to beat the other fellow. 'And I've raised lots of prize-winning birds that way. I never cared to show them myself, but sold them to breeders who showed them'."

29. KEYES CHADWICK. Whistler hen from 1900-1910 with painted eyes. The smallest of Chadwick's decoys thus far recorded.

It is difficult today to realize how popular the raising of fancy poultry was in years gone by. A sizable group of Island men engaged in it, and Chadwick was the most successful by far. Four male birds of his rearing won first honors in National shows at Madison Square Garden. One of his White Plymouth cockerels became grand champion of solid colored birds over almost 6000 entries.

The same sharp eye for judging line and volume was useful in other ways. He was once asked to estimate the combined weight of three enormous hogs which belonged to a neighbor. Other men placed bets on whether or not he could come within twelve pounds, but he came within eight. The total weight involved was well over 1500 pounds.

Recently a man who shot many a rabbit in company with him asked me, "Did you ever hear Keyes Chadwick say an unkind word about anyone?" No. But then I met him only near the end of a very long life. His contemporaries knew him to be a man of unquestionable integrity, a kind man, a gentleman.

His wife was childless and the two were very close. When he rose in the predawn hours to go hunting, she was up to fix him a hot breakfast, and one observer claims that Chadwick never ate a bite of food unless it was in his own house. It made no difference how long he was out gunning or who had box lunches; he would wait to eat with his wife. Their friends remember that the Chadwicks treated each other

30. KEYES CHADWICK. A great pair of American mergansers from Chadwick's middle period, 1910-1930. They are from a set of a dozen birds of this species which was carved for Capt. Norman Benson of Lambert's Cover in trade for a rabbit dog. All paint has been removed.

31. KEYES CHADWICK. A detailed portrait of the majestic American merganser hen in Plate 30.

32. KEYES CHADWICK. *Goldeneye (whistler) drake, overpainted by a later hand. 1920-1930.*

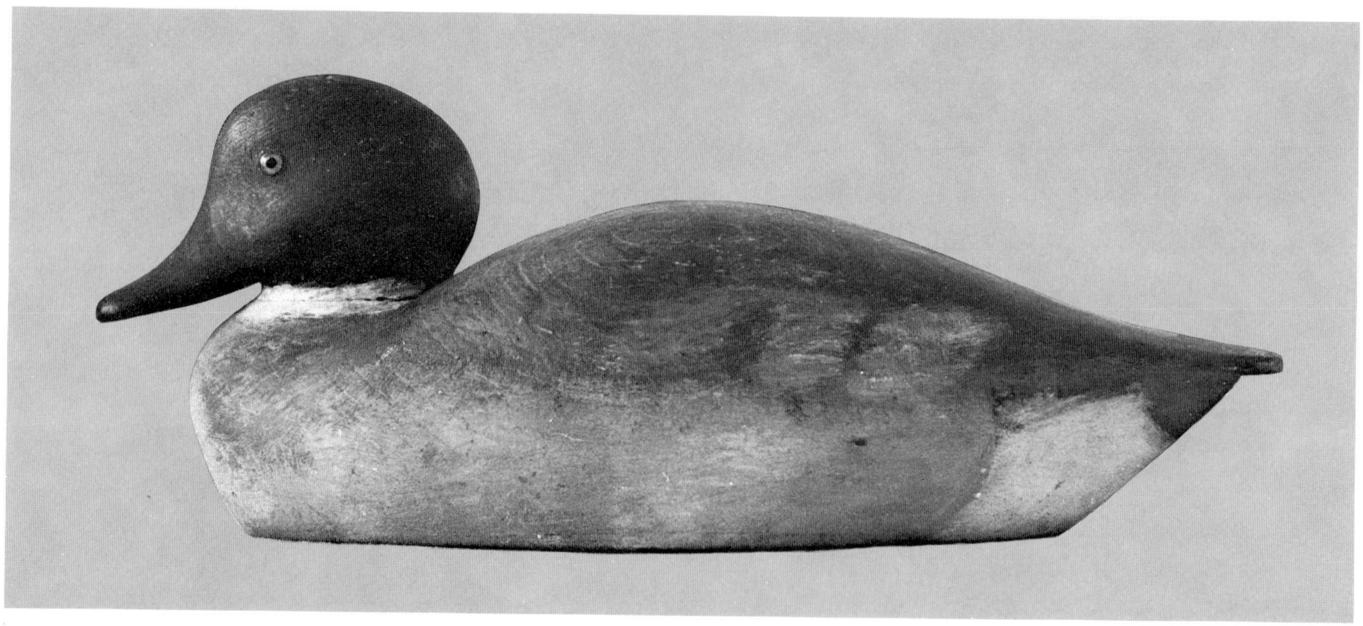

33. KEYES CHADWICK. *A whistler hen at rest. Overpainted by another hand. 1930-1940.*

34. KEYES CHADWICK. American goldeneye, or whistler. The paint is original. Chadwick had separate cardboard patterns for the white wing-markings which were painstakingly traced on the block. 1940-1950.

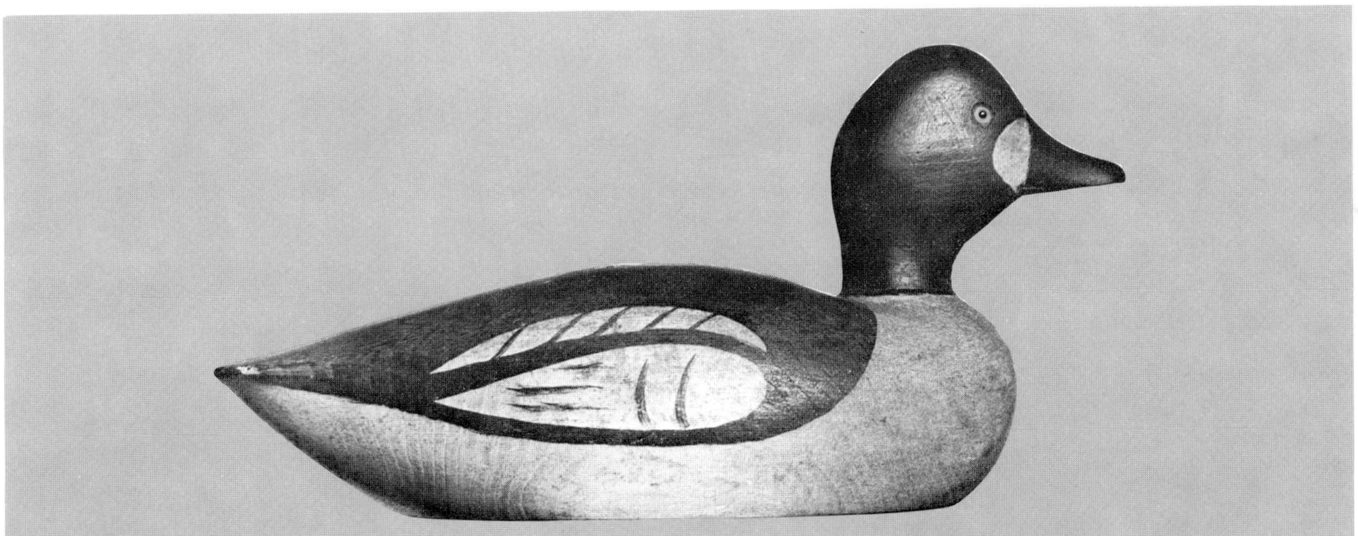

35. KEYES CHADWICK. One of the rarest of working decoys, a Barrow's goldeneye male in original paint. Not only does the carving of the head catch the subtle characteristics of this species, but the paint of the head is the proper blue-purple-black. 1940-1950.

36. KEYES CHADWICK. A ruffed goldeneye drake with worn, original paint. 1930-1940.

with total consideration always, living very much for each other yet carefully respecting one another's privacy.

At a time when most men are content to observe life from a porch rocker he still went gunning on occasion, and a small paragraph in the Vineyard Gazette for January 12, 1950, reported that "H.K. Chadwick of Oak Bluffs, who has passed his 85th birthday, took his first jeep ride the other day with Joe Amaral. The quest of the two was an eventual rabbit stew, and Keyes got his rabbit, too." All of his life he much preferred rabbiting to wildfowling, and kept as many as six beagles for the sport.

He stopped making decoys in 1951 and ceased caring very much about anything after Mrs. Chadwick's death in 1953, except that he continued to be gentlemanly to the rest of the human race. He had a habit of giving little baskets carved from peach pits to people who visited him, if he liked them. The tiny basket was a sort of trademark and a proud one at that. When he gave one to a visiting interviewer (he was then 85) he suggested that the young man give it to a girl and pass it off as his own work. She would know it took patience to whittle such an item, he said, ". . . and girls like a man who is patient."

There is one consistent and sad note which runs through the interviews given by Keyes Chadwick in his last years, and it appears in certain of his letters as well. He made it clear that he thought very little of

37. KEYES CHADWICK. Redhead drake in original paint from one of Chadwick's favorite patterns. 1930-1950. Collection of Duke's County Historical Society.

38. KEYES CHADWICK. Detail of the male redhead in Plate 37. Perhaps his most often-repeated head.

39. KEYES CHADWICK. Red-breasted merganser pair. These are from the last batch of fully-painted Chadwick decoys. 1940-1950.

his carving ability and felt that it had no importance whatever in his life, which he considered to have been rather wasted. "Ben Smith and Elmer Crowell now," he said, "they were artists with the knife and with paint." He himself would have found fulfillment in raising champion chickens.

In the final three years spent in a nursing home his mind and memory remained clear, but his then thin body (he had been described by a reporter a few years earlier as "short, heavy-set") was worn out. There was a small shelf above the foot of his bed, where a shoe-box contained some of his peachstone baskets, a few samples of his lovely, formal handwriting, yellowed clippings about his poultry-prize triumphs, and some keepsakes of his wife. He died on September 16, 1958, aged ninety-three, and was buried next to his wife, a West Tisbury girl, in the cemetery of that town.

40. KEYES CHADWICK. A brant from a set of twelve made to order for one man.

41. MAX CHAMBERLAIN. Bluebill hen (above) and drake. Worn, original paint.

42. MAX CHAMBERLAIN. The bluebill hen of Plate 41.

Max Chamberlain

All that I can discover of Chamberlain is that he was a hunting crony of Cooper Gilkes, of Oak Bluffs, who ran several gunning camps around 1900. I have seen a half-dozen of his birds, all bluebills. The wing-tips and tail-feathers are incised where they cross the back; the eyes are shallow-set, giving a pop-eyed and comical look to his decoys.

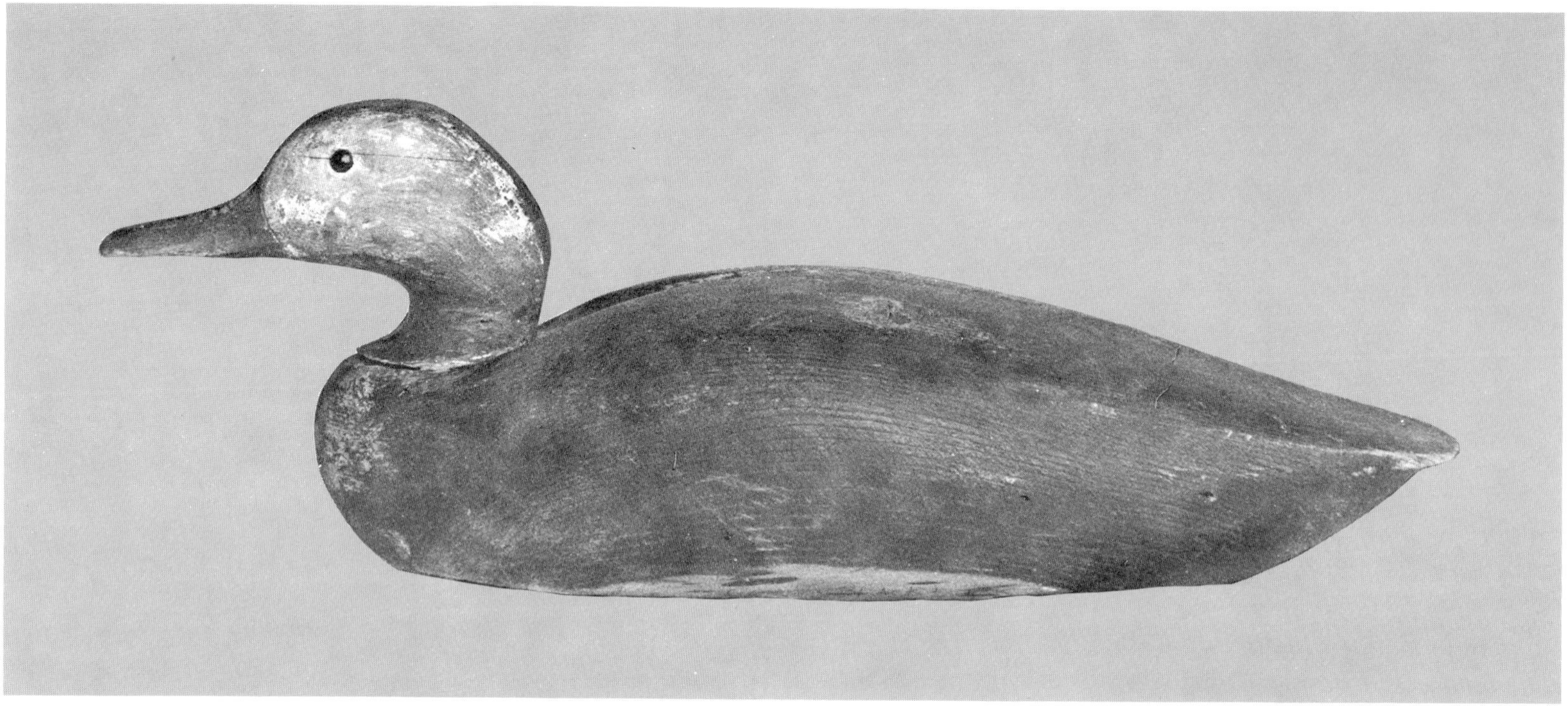

43. WILLIAM E. DUGAN. A Group 1 (see text) black duck. Dugan's earliest carving style.

William E. Dugan, Sr. 1892–1955

Any admirer of decoys who is familiar with the Chadwick style is sooner or later going to come across the work of W.E. Dugan and will be surprised and, very likely, puzzled. Rightly so.

Because Dugan borrowed certain patterns of Keyes Chadwick and made a number of decoys from them. And he had the natural talent and craft to make some copies that are near-perfect. With a few exceptions all of the decoys he owned were branded by him with the legend W.E. DUGAN, but this is not to say that he made all of the decoys he branded. In a letter to a collector-friend, Keyes Chadwick wrote: "I sold – and Dugan 300 decoys and they divided them, if dividing is what it might be termed. Dugan got the best of them . . ." (Dukes County Historical Society archives.) All now carry the brand of Dugan. Since he copied not only the carving but the paint pattern as well, questions of identification are bound to arise.

The imitations are the more difficult to understand when one realizes that Dugan was artistically talented to a high degree, and carved a quantity of completely original works which included decoys, ornamental birds and miniatures, and large high-relief sculpture of considerable power and beauty.

William E. Dugan (rhymes with you-*can*; the name is not Irish, it derives from the Portuguese phonetic equivalent) was born in Edgartown and moved to Vineyard Haven as a boy. After service in the Navy during World War I he returned to Vineyard Haven and worked briefly at odd jobs. He seems to have possessed a born ability to draw fluidly and well and this led him to try his hand at designing houses. According to the man who worked as his foreman for years, he had no formal training in architecture and had not served an apprenticeship as carpenter. Nevertheless he could almost instantly translate the wants of would-be home owners into drawings that gave attractive shape to their dreams.

He gathered together a competent group of carpenters and began a contracting business that eventually produced some of the handsomest and best-built Island houses of his time.

Like many creative people of the past and present, he had a poor head for business. His scrupu-

44. WILLIAM E. DUGAN. A Group 1 Canada goose, the entire concept is Dugan's as far as can be determined at this time.

45. WILLIAM E. DUGAN. A closer view of the Canada goose in Plate 44. One of Dugan's earliest carving efforts and totally unlike the geese he later turned out in quantity (Plate 48).

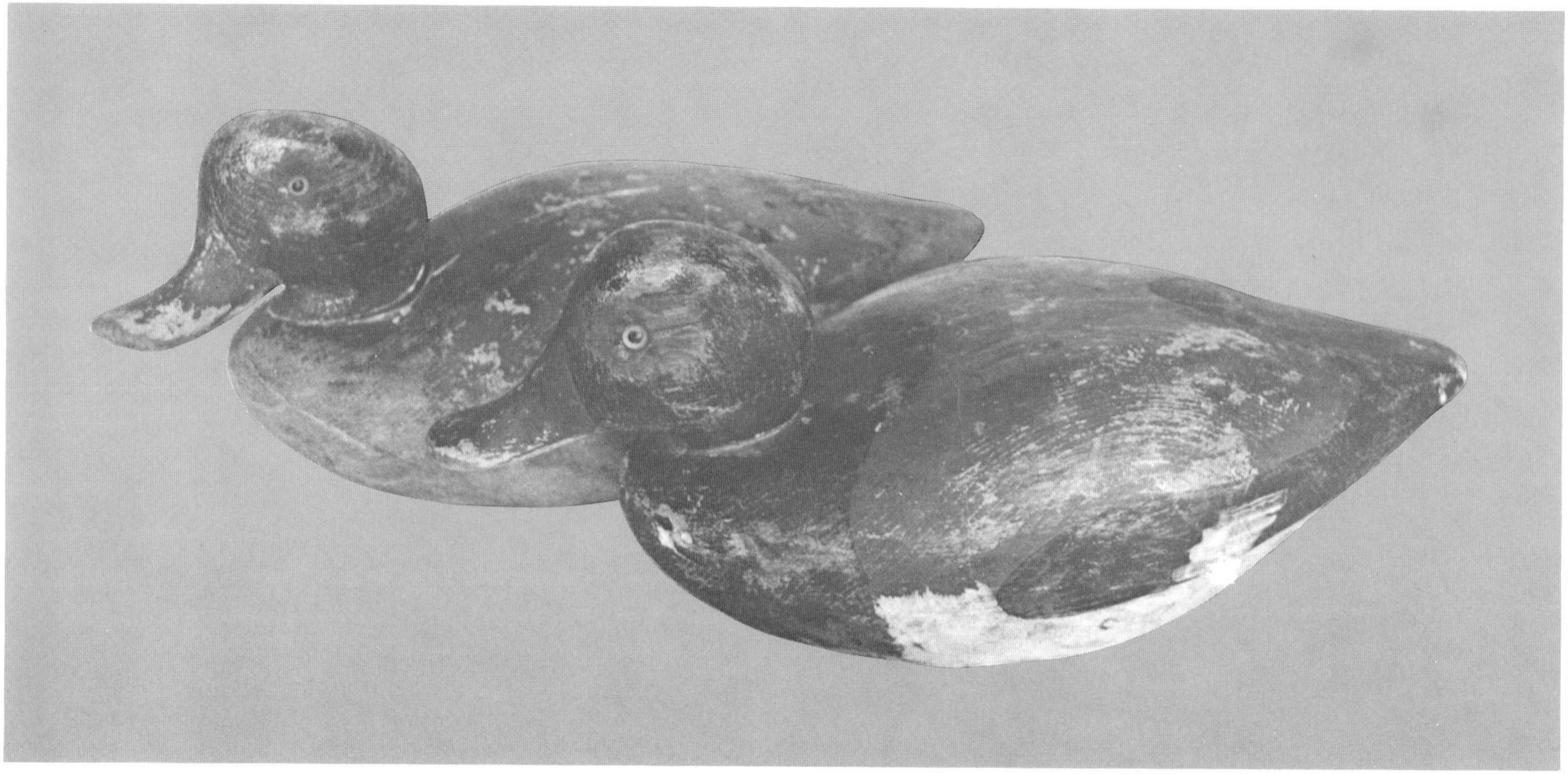

46. WILLIAM E. DUGAN. Pair of Group 2 bluebills. These are crude copies of Chadwick's work, turned out in Dugan's shop by his foreman, working from a Dugan pattern.

lous interpretation of moral honesty made him loathe to try to make any profit above the cost of his labor and materials—he called such profits "blood money"—and as a result, over a period of years, slowly but certainly went into debt to his suppliers. Finally he was forced to declare bankruptcy, which seems to have dealt a mortal blow to his desire to continue as a contractor-designer even though everyone he ever worked for held him in the highest esteem.

His family life—he had three sons— ended in a separation from his wife when the boys were small and his later experiences with women are said to have resulted in the loss of most of his tangible property. A teetotaler until he was past forty years of age, he became a hard drinker. His last years were lonely except for his hunting in the fall and winter when he was again with old friends sharing hours in the field and marsh.

I asked a close friend of Bill Dugan, a man who knew him well and remembers him fondly, about what connection Dugan may have had with Chadwick in his lifetime. The friend, now a clear-eyed, active

47. WILLIAM E. DUGAN. A Group 3 merganser. This bird is a close copy of the only Chadwick merganser I have seen with the comb thusly carved. Horribly overpainted by a later hand.

man of eighty years replied that there was no close connection at all. He said that Dugan felt that the Chadwick decoys were the best and were worthy of imitation. Simple as that. We know, though, that he visited Keyes' shop more than once. He did, after all, own many Chadwicks, and it is known that one of his regular hunting companions was Elmer Chadwick, the brother of Keyes.

On April 29, 1955, in district court, he entered a plea of not guilty to charges of drunkenness and driving "under the influence," as it is called in Commonwealth law procedure. A few days later he took his own life. The Vineyard Gazette of May 6, 1955, reported:

"William E. Dugan, Sr., Vineyard Haven contractor, designer and builder, ended his life early Monday morning by shooting, in the cottage he occupied on Beach Road. Mr. Dugan had built many of the best-known and most attractive houses and his productive years were marked by great ability and accomplishment . . . He lacked a few days of being 63 years of age and had been in poor health for some time."

The Dugan decoys may be divided into three distinct groups: In the first—and these are rare—are the birds entirely in his own style and by his own hand (Plates 43-45); In the second are the decoys manufactured for him in his own milling shop by two of his carpenters, working from patterns drawn by

48. WILLIAM E. DUGAN. A Group 2 Canada goose carved for Dugan by his shop foreman who worked from a pattern Dugan took from a goose made by Jim Look (see text pages 79-81).

49. WILLIAM E. DUGAN. Close-up of the goose in Plate 48.

50. WILLIAM E. DUGAN. Top, a Group 3 Dugan redhead drake. Bottom, the Chadwick original.

Dugan (Plates 46-49). They are generally imitative of Chadwick but are a far remove from his quality. In the third group are the excellent Chadwick imitations, some of which would, I think, fool that master himself (Plates 47, 50). I believe that these last are Dugan's own handiwork in every detail and must date from the years after he left the contracting business.

For comparative purposes, the Dugan-Chadwick differences (as apply to Group 3 redheads only, Plate 50) are listed here:

1. The Chadwick bill is narrower at midpoint.
2. All of the Chadwick bills have a slight drop below the tip of the bill. No drop in the Dugans.
3. Chadwick painted the tip of the bill (black) top and bottom—Dugan, top only.
4. The bottom of a Dugan decoy is smaller in area.
5. Dugan backs have a somewhat flat profile. The Chadwicks have quite a high curve.
6. The Dugan weight is 1-5/8 inch diameter; the Chadwick weight is 1¾ inch diameter.

The W.E. DUGAN brand was burned into the bottoms of the duck decoys and into the upper backs of the Canada goose blocks. It exists in two slightly varying forms because, for some reason, there was an iron branding iron, and a bronze branding "iron". They appear to have been used interchangeably.

51. RUSSELL HANCOCK. Three roughly finished but convincing lesser yellowlegs, stained (rather than painted) a dull brown for the Fall shooting. The attribution to Hancock is tentative.

Russell Hancock 1842–1931

He was born in Chilmark on the family farm at Quenames, a very beautiful land which abuts the Atlantic Ocean and the shores of several ponds. The farm is still being worked and is still owned by Hancocks.

As did many young men of the Vineyard of his generation, Russell Hancock went whaling, pursuing the great mammals for some twelve years, shipping as mate on his last voyages. He left the sea to return to Chilmark and work at the trade of carpenter and farmer, and fish the ponds commercially. Among other offices, that of county commissioner was one he held for a number of years.

For his own use, he made some small, solid body, quite primitive duck decoys. All that I have seen have been red-breasted mergansers, thinly painted in a rather offhand manner. The white cedar bodies are flat bottomed, with pine heads joined to the body by a screw driven through a hole bored up from the decoy base. Several of his ducks have a large, simple H chiseled into the bottom with three strokes of the tool.

Quenames was one of the great, fecund wildfowling areas, a natural flyway of the south shore for all manner of birds, and if few of Hancock's decoys exist today, I venture to suggest that it is because the pass-shooting (wing-shooting from blinds along the known routes of flying ducks) was so productive that tolling was unnecessary.

The pair of shorebirds shown here (Plate 51) cannot positively be attributed to Russell Hancock, but he gunned over them and they are primitive enough to be his work. It is only a possibility. Herbert C. Hancock, the son of Russell, told me (he was then over eighty, a strikingly handsome old man) that he and his brother once killed two barrels of golden plover in the salt meadows of Quenames. They were shipped to Boston where they brought 50 cents a pair. One of the few places in the world where this species may still be seen today with any regularity is on the same Quenames meadows.

52. RUSSELL HANCOCK. *Red-breasted merganser hen.*

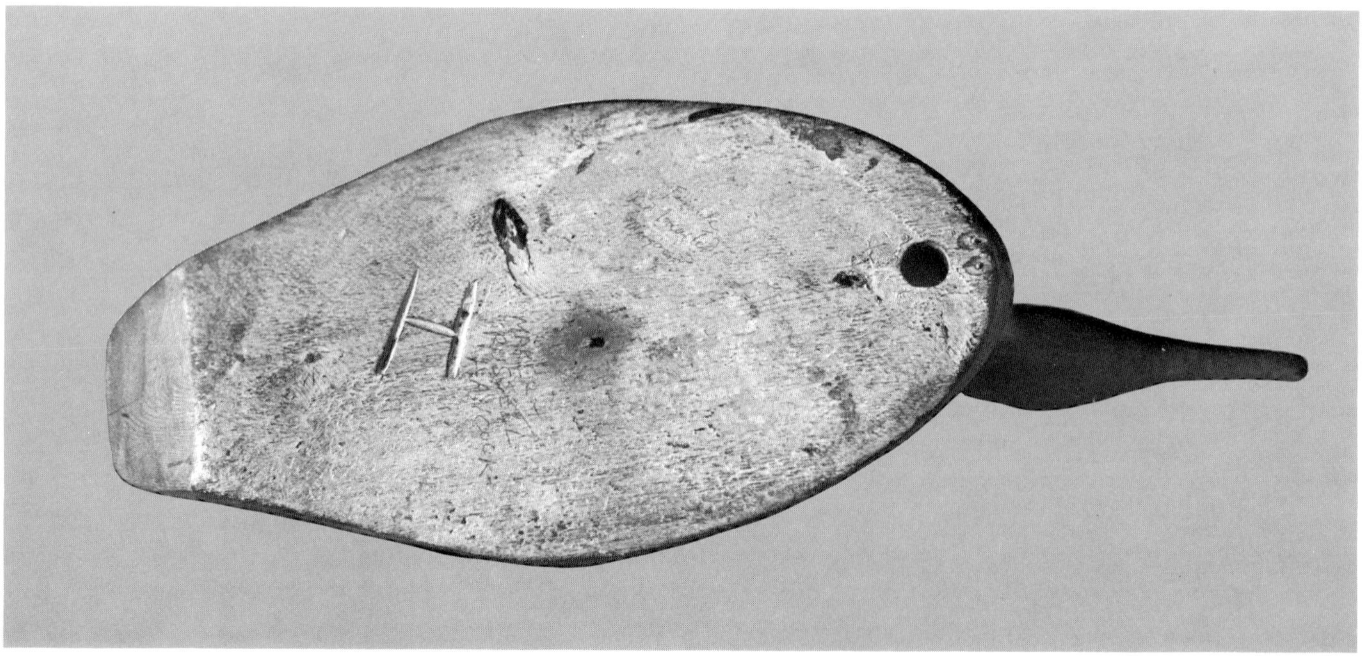

53. RUSSELL HANCOCK. *Bottom view of the shelldrake in Plate 52.*

54. RUSSELL HANCOCK. Pair of Red-breasted mergansers.

Neither of Russell Hancock's sons made decoys though both were life-long gunners. His grandson Hariph, however, is known to have made at least one set of merganser decoys for which the pattern still exists. The great-grandson of Russell, Herbert Hancock, now a building contractor in Chilmark, is a more gifted carver than his ancestor. His ornamental shore birds grace a number of Island shops.

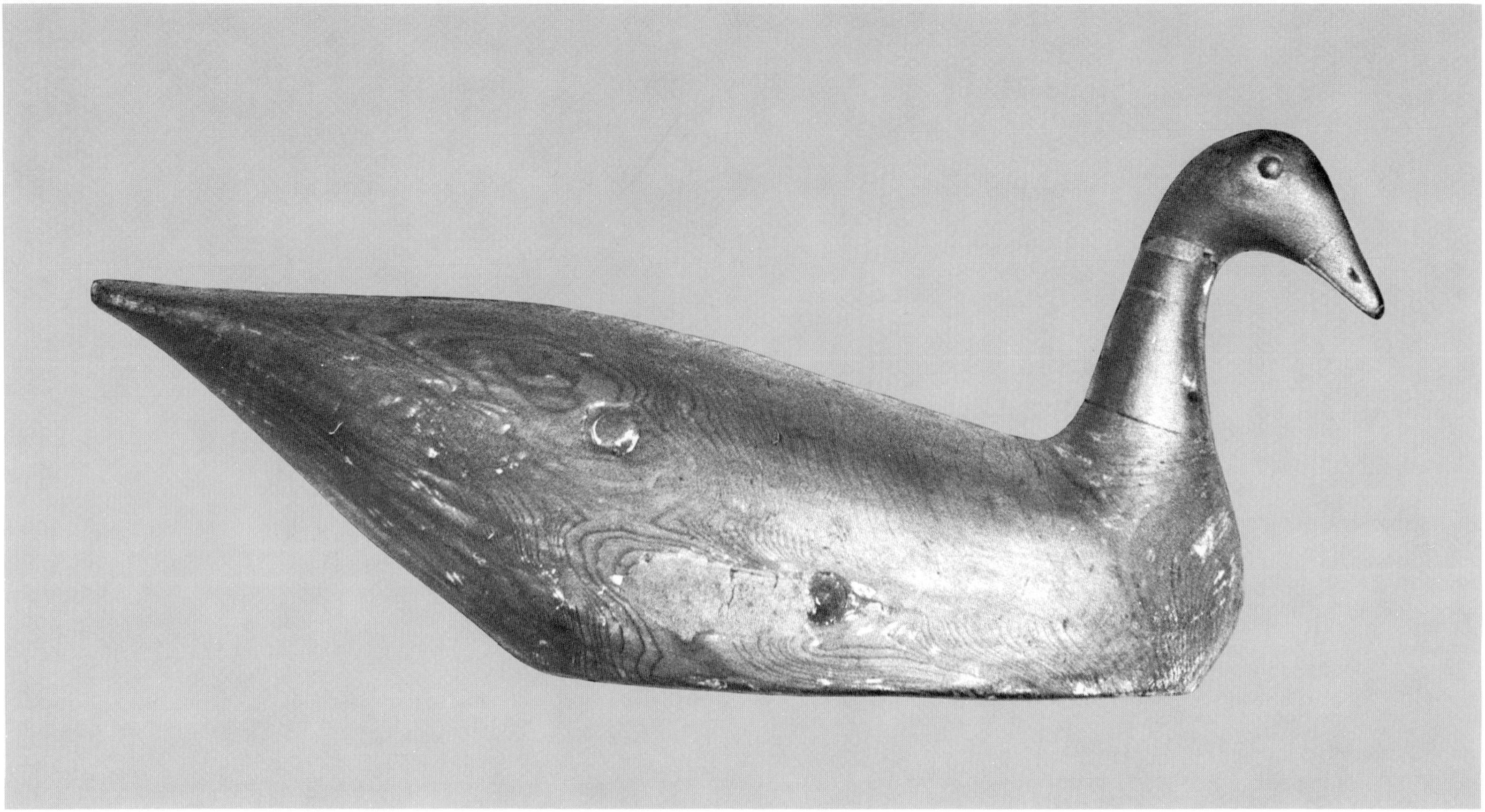

55. JACKSON-CROWELL-ROBERTS. Brant. All three men worked from the pattern for this bird. I believe this specimen is by Hiram Crowell. Eyes are round-head brass tacks. All paint removed.

Robert L. Jackson, Jr. 1877–1953
Hiram Crowell 1885–1953
Manuel Swartz Roberts 1881–1961

These three men, all of Edgartown, are presented together because they all took part in the creation of a particular rig of brant. They assembled on a number of evenings during the banner brant years of 1920-30 and carved these decoys, all working from the same pattern.

Because a large number of decoys were necessary to toll brant, the trio made well over one hundred of the little geese. Their decoys were pooled with those of other hunters to create a huge stool to be set over the then-flourishing eel-grass beds in the waters off Edgartown harbor and in Nantucket Sound. Pairs of hunters took turns going out to lie in a boat up-wind of the decoys, each pair shooting a limit before returning to shore.

When I first spoke to Captain Jackson many years ago, there were still over twenty-five of these brant in his garage, along with some whistlers carved entirely by him. Most of the others, he said, had "gone for kindling," and he meant just that.

The lives of all three men were tied to the sea in one way or another. Jackson was a successful commercial fisherman, for years the captain of a dragger. One of his early commands was shelled to the bottom of Georges Bank by a German submarine in World War I, with the men left to row 140 miles to land.

Crowell and Roberts were both boat-builders until Crowell opened one of the first gas stations in Edgartown. As a carver Crowell remains somewhat of an enigma. I suspect, on the basis of an exquisite widgeon that I saw years ago, that he was one of the more talented Vineyard carvers. The widgeon has since disappeared, but it was direct kin to the best of the rig of brant produced by the triumvirate and was beautifully painted as well.

56. ROBERT L. JACKSON, JR. Brant. Entirely by Jackson. He carved a number of them on the long runs to Fulton Fish Market, New York, from the off-shore grounds. Original paint.

Manuel Swartz, as he was known to all (the Roberts was added to his name at the insistence of his second wife; it was her surname), said that these brant were the only working decoys he had a part in. He did make a few decorative birds in his last years. His legendary quality as a man is memorialized in the following words on a bronze tablet fixed to the front of his old shop:

> For many years this building was the shop of
> Manuel Swartz Roberts, boat builder.
> From its doors came catboats, beach boats, fishing boats,
> all staunch, all honest, all seaworthy, all partaking of the
> nature of their builder.
> Through its doors also came his many friends and in
> particular the young sailors of the town to gab, to learn the
> use of tools, and to draw inspiration from his homely philosophy,
> keen wit, and never failing kindness. Manuel was never too
> busy to give of his time and heart. Truly he was also a
> builder of men.

57. ROBERT L. JACKSON, JR. *Goldeneye drake. The drawknife marks are clearly visible.*

58. JAMES LOOK. Four redhead drakes ready for action. No two are exactly alike but all have Look's characteristic indentation of the upper back, and bear the sharp chin line as well. Original paint.

James Look 1862–1926

Jim Look—no one seems ever to have called him James—lived all of his years in the town of his birth, West Tisbury. His home stood, and still stands, at the head of one of the numerous covers of Tisbury Great Pond and it was there as a boy that he began to fashion small boats for his own use. He didn't have a formal apprenticeship in the trade but he gradually became known far beyond the Vineyard as a boat-building craftsman, and to his shop on the shore of Tiah's Cove came orders for skiffs, sailing craft, and work boats for use in the trap-fishing industry which then flourished in Island waters. As an example of his ability, of his superior eye, there is the legend of a time when Look went to Oak Bluffs where a man was building a boat and needed help to bore the shaft-hole through the keel. Jim tapped a nail into the end of the keel where he wanted the shaft to come out, then used a hand auger from inside the boat, working entirely by eye and instinct. When the auger came through a distance of six feet, it pushed out the nail.

Jim Look was also a life-long fisherman of the salt ponds, seining herring and perch, bottom-fishing and eeling, but he is perhaps best remembered for his gunning. An article printed at the time of his death reported: "Always very active and gifted with an abnormal instinct for locating game, he became one of the most skillful wild-fowl hunters that the Island ever had, and it has been said of him that nowhere in the state could a man be found who was his equal at this sport."

For many years in the latter part of his life he turned this talent to profit as a wildfowling guide. His considerable shore holdings on Tisbury Great Pond, and adjoining or nearby properties which he leased, had sites for numerous gunning blinds. During the open season off-Island sportsmen came for varying lengths of time to board in the roomy Look house and were taken to their alloted blinds in the cold, pre-dawn hours each morning.

But the main attraction for visiting hunters was the goose camp, where live flyers were used to lure the migrating Canada geese. The work involved in this enterprise was considerable because the care of

59. JAMES LOOK. The redheads of Plate 58. Today the species is a rarity. In Jim Look's time the flocks on a single Vineyard pond could number in the thousands.

the many mated pairs and their goslings of each season was a year-round responsibility. One of today's residents of West Tisbury, George Magnuson, came to the Island as a youth to be employed full time by Jim Look. He held the job for eight years, until Look's death, and remembers him now, fifty-odd years later with something akin to reverence. "Hardly a day goes by" he says, "that I don't recall Jim and how much he meant to me."

Magnuson says that Look made as many as 200 decoys. He *used* up to 600 a season but most of them were factory decoys brought here by the various city nimrods who came to gun under the Look guidance. I have seen only redheads and bluebills by Jim Look, and one Canada goose in the collection of Dr. George Starr of West Duxbury. I am not aware that he carved other species.

60. JAMES LOOK. A greater scaup (bluebill) drake from Look's early style.

Decoy bodies were cut from white cedar planks 20' x 12" x 4". They were roughly sawed out, then worked down with a hand axe, a rasp, and finally sandpaper. Heads were pine, joined to the body by a screw from below, the hole then corked in the Chadwick manner.

The jaunty heads of the Look duck decoys are almost interchangeable from early period to late. The early bodies are less chunky than the late but all of them have the deeply hollowed upper back. Weights can be a rough circle or rectangle of lead, nailed on. The paint patterns are simplified and subdued with no attempt at blending. Usually there are several coats of paint, all done by Look, because in a good redhead year a bluebill might be repainted for that species only to become a bluebill again a few seasons later. Eyes are merely painted circles.

61. MATTHEW MAYHEW. A pair of small, red-breasted merganser hens with narrow heads and delicate necks. The upper example is in original paint; the lower has had all paint removed and the neck repaired. The eyes are carved in the manner of Mayhew's neighbor, Frank Adams.

Matthew H. Mayhew 1847–1940

He was a carriage-painter, a job he learned during a decade spent off-Island in his youth. He returned to West Tisbury, the town of his birth, to practice this trade for as long as the horse-and-buggy age lasted, and to build a reputation as the best of his line. When the automobile came into vogue he easily switched to painting them and continued to do so for many years.

Since his business tended to be seasonal, Mayhew hunted for market and fished commercially in the local waters. He was noted as a hunter of foxes which at the time existed in quantity on Martha's Vineyard. A state bounty was offered for their heads.

When he died at the age of 93 the Vineyard Gazette noted that he had been "Ill only a short time and had been able to attend to all the duties of the house which he had occupied alone for many years. Short and wiry, he walked with springy step and his eyes were keen, his hearing perfect, and his voice with little of the quaver of age."

Matthew Mayhew is presented here as a decoy maker on the basis of one set of perhaps twelve red-breasted mergansers all of which were obtained directly from him by a man to whom he said "I made these." If not for this statement it would be logical to assign the decoys to the ouvre of Frank Adams, the friend, gunning companion and neighbor of Mayhew. Together they leased a stretch of Tisbury Great Pond shore for their duck hunting. Stylistically these small, solid shelldrakes are faithful counterparts of similar birds by Adams. It is the neat, rigidly defined paint pattern which differentiates the work from Adams' own; the straight, clean lines of a carriage painter.

62. WILLIAM H. MAYHEW. Goldeneye drake. This small, hollow character wears his original coat of paint.

William H. Mayhew 1857–1944

The Mayhew name is an ancient one in Vineyard history, a family tree with many branches and twigs. Matthew Mayhew of West Tisbury (page 83) and Willie Mayhew of Edgartown (and another Willie Mayhew—this one a legendary fisherman and hunter of Chilmark) were distant kin and were active (along with many other Mayhews) in roughly the same period of tine.

The parents of William H. Mayhew, lured as were many other Islanders by the golden promise of the great West, left Martha's Vineyard to settle in California. Their son was born there but when he was a small boy they returned to the Island, and Willie Mayhew spent his life in Edgartown.

He is recalled by those with long memories as the proprietor of a hardware store which carried a large stock of things common and rare, where items were stored in every conceivable space, floor to ceiling, cellar to attic, and where in winter the local men gathered to gossip by the stove. He served as Town Clerk for several terms and remained a bachelor for all of his 87 years.

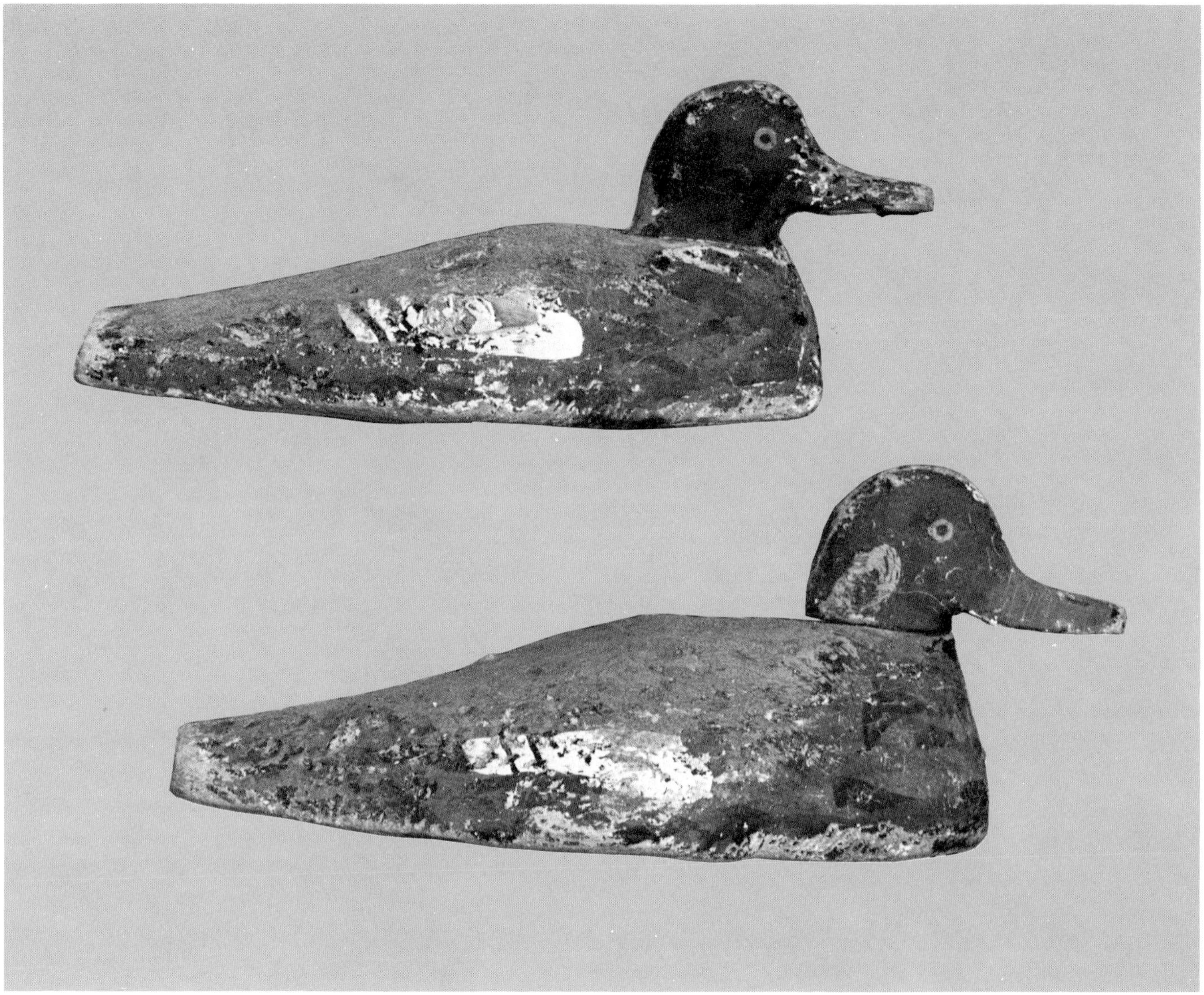

63. ALLEN NORTON. *Pair of bluebill hens.*

Allen Norton 1831–1911
Winthrop B. Norton 1905–

Of the known decoy carvers represented in this book, only Winthrop Norton of Edgartown is alive today. There are several Island craftsmen making wooden decoys here but their birds are meant for the mantel, not the marshes, and are not within the scope of this book. Joe Thomas, also of Edgartown, who died in 1970 aged 80 years, turned out a broad variety of decorative bird carvings for years. He did make one stool of twelve brant for gunning use back in the forties, but since the rig saw very little action and is still intact and in the possession of the man who ordered it, inclusion of them here would serve little purpose. They are very similar to the brant made by Capt. Robert Jackson (page 76).

Norton, an Island architect (the Oak Bluffs town-office building is one of his recent works) made a string of one hundred bluebills and twelve Canada geese in 1930. They are in the direct tradition of the older decoys; made to be used by the maker who had his own unique vision of the birds they represent. They have turned up in collections as far away as New Jersey as the indirect result of a natural disaster—the hurricane of September, 1938. At that time they were stored in a small gunning camp near the ocean beach on the Norton property. When the storm blasted the camp, and the accompanying abnormal tides scattered the decoys far and wide, Norton was able to recover very few of them. A lot of other decoys changed hands as a result of that same hurricane; William E. Dugan (page 61) lost hundreds of them.

The Norton land, a rich variety of field, forest, great pond, dunes and ocean beach, has been in the family for many generations. The grandfather of Winthrop, Allen Norton, made his living from it as farmer and fisherman and he too made decoys for his own use. They are as simply conceived as those of his grandson. The three illustrated here are all that still remain in family hands.

When Winthrop's father, Frank L. Norton, inherited the property he was a building contractor, but he lived on the farm and was an avid gunner. He never made a decoy but he was a friend of Keyes Chadwick and owned a number of that master's carvings—some so early that they are unweighted, others from

64. WINTHROP B. NORTON. Bluebill drake. Fashioned with bandsaw and lathe.

Chadwick's last working years. All shelldrakes, by the way. To Frank Norton, as to so many other men of his generation, the whisper of wings in the marsh was a lure transcending all others. In the drugged hours of his terminal illness he murmured of birds in flight, he swung an imaginary shotgun. His son's words might be the epitaph of most wildfowlers: "His mind was on the ducks to the last."

65. ALLEN NORTON. Species not identifiable, paint weathered away.

66. WINTHROP B. NORTON. Canada goose. Another lathe-turned decoy, in three pieces. The paint has not been stripped; it has weathered away.

67. BENJAMIN PEASE. Redhead drake. The eyes are round-headed iron tacks. Original paint.

68. BENJAMIN PEASE. The redhead of Plate 67.

Benjamin Warren Pease 1866–1938

Some years ago I paid a visit to Manuel Swartz Roberts of Edgartown when he was still working in his boat-building shop hard by the Chappaquiddick ferry slip. We discussed decoys and he fished a fine old redhead out of his loft and said it had been made by "Ben Pease of Chappy." At that point he was the only person I had questioned who had any idea that Pease made decoys. Thanks to Manuel Swartz, the carving style, expert and assured, of Benjamin Pease can now be readily identified.

He was born on the family farm called Tom's Neck on Chappaquiddick, a lovely, rolling land very close to the sea. The north border of the farm is on the broad expanse of Cape Pogue Bay. The east side is separated from the dunes that overlook Nantucket Sound by a long, narrow herring creek connecting Cape Pogue Pond with Pocha Pond.

When he was a young man (and after a year on the west coast where for a time he was a cable-car conductor in San Francisco), Pease was apprenticed to an Oak Bluffs carpenter. He married and settled in Oak Bluffs and lived there for about fifteen years. Not long ago I spoke to his last surviving child, an attractive, chatty woman now over eighty years herself, and she remembers that the family moved to the ancestral land on Chappaquiddick when she was thirteen. What is more pertinent to this account is that she recalls once being in her father's shop during her Oak Bluffs childhood when her father and Keyes Chadwick were both carving duck decoys. The date would have been around 1904.

After the move to the farm, Ben Pease spent an active, hard-working life as builder, farmer, offshore fisherman and seiner of herring, and was a fish buyer, selling to Fulton Market in New York. He also served his town as selectman and assessor and managed to find time to go gunning in season.

When he died, the Vineyard Gazette noted that "He often said that he lived in the most beautiful place in the world and that he felt himself fortunate to have been so privileged." A few weeks later the Gazette published a rather ornate tribute to Pease from one Electus Litchfield which is quoted here in part:

69. BENJAMIN PEASE. Bluebill drake. Original paint.

"Mr. Pease was in so many ways the ideal good neighbor. His character had something of the sweetness and beauty that we associate with the loveliest of inanimate things; something of that quality that there is in a Mayflower and in the butterfly weed; in the beauty and comfort of a great tree on a summer's day. He was the first citizen—the first gentleman of Chappaquiddick, and it will never be quite the same place without him."

Another sort of appreciation for his skills was indicated when, during the few days between his death and his burial, someone thought enough of his decoys to steal them from his barn.

* * * * *

Benjamin D. Smith, Keyes Chadwick, and Benjamin Pease were within a year of the same age. In their early thirties all were living in Oak Bluffs and working as carpenters and it is easy to suspect that they may have hunted together now and then. We know that Chadwick learned decoy carving from Smith; it is possible that Pease then learned from Chadwick.

An examination of the Pease redhead (Plate 67) shows that he employed the Chadwick flush-poured weight. The head is joined to the body in the Chadwick manner and all parts are finished smoothly,

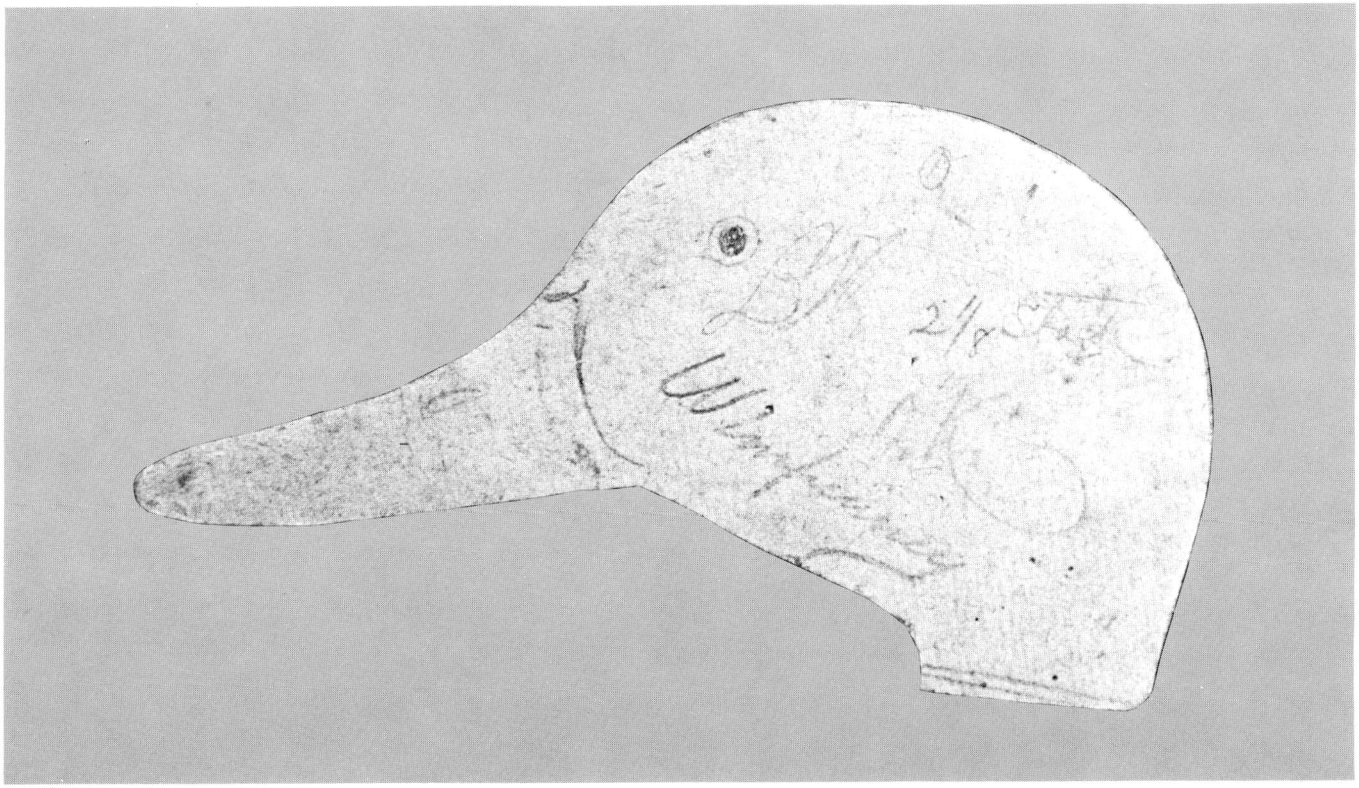

70. A mystery. This pattern, cut by Keyes Chadwick and found with his other patterns, if for (or from) the bluebill of Plate 69, by Ben Pease. Why is the name Wimpenny (page 117) written on it?

with care. It is not the work of a man who just happened to carve a few ducks; it is handled professionally all the way.

Beyond these merely technical points, the Pease conception of a duck is entirely original. Unlike Smith or Chadwick the Pease decoy carries its tail high and has greater breadth of bottom. The neck is somewhat thin, yet strong, and the well-shaped head is slightly flattened along the curving crown. Eyes are metal studs.

Stylistically, the Pease bluebill (Plate 69), while smaller, is very similar to the redhead in all points except that the bluebill is not weighted. The paint pattern of both birds is original and is simple, flat, and effective, with no attempt at blending.

Materials are the traditional white cedar bodies and pine heads. The easy, firm, perfectly symmetrical carving of the few Pease decoys I have seen thus far suggests that they are only a fraction of a large rig. Both bluebills and redheads decoyed best to a big stool, the more the better.

71. FRANK RICHARDSON. Brant. Before painting this graceful sculpture, Richardson neatly fitted a section of wood into the tail to replace a flaw in the block.

Frank Richardson

Biographical information on this man is not only elusive it is almost non-existent. He came to Martha's Vineyard from Maine, possibly by way of Nova Scotia, and established himself as a fish-buyer in Edgartown. He was not young when he arrived here and there is no subsequent report of his death in the town records, so presumably he departed for parts unknown after a residency in Edgartown of perhaps ten years.

He was a bachelor and lived near Anthier's, a sizable tidal pond shared by Edgartown and Oak Bluffs. With no definite proof I tentatively assign his presence here to 1920-1930. My only good reason for doing so is that he made a number of brant decoys and those were good brant years.

All of Richardson's decoys that I have seen were given or sold to an Edgartown gunner, and it is from this gunner, now deceased, that the examples of Richardson's work illustrated here were obtained. The single specimen of his art in the decoy collection of the Massachusetts Historical Society came from the same source.

He is known to have gunned in the same area as Ben Smith, on Anthier's Pond, and there is clearly a connection between Smith and Richardson in certain decoys of both men. There is a Smith shelldrake (Plate 79) which has a replacement bill carved by Richardson—it has his distinctive way of indicating tiny teeth—and the Smith cock shelldrake paint pattern is on several birds owned (if not made) by Richardson.

Richardson was a craftsman with wood. Almost every duck or brant has some variation, some repair, some expert mortising done where he was dissatisfied with the block. In his hollow birds the hollowing is different every time; oval, round, rectangular. Weights are different too, but tend to be flattened ovals. In the three or four solid, round-bottom birds I've seen the lead is countersunk in a narrow strip. Glass eyes tend to protrude slightly but are high-quality glass.

There is such variety in Richardson's work that I sometimes feel he may have taken the decoys of

72. FRANK RICHARDSON. Red-breasted merganser. A strong, solid, round-bottomed female shelldrake. The eyes are glass and the mandibles have the tiny serrations of all of Richardson's mergansers. The weight is a small, oval-ended, counter-sunk strip of lead.

other men (possibly Maine?) and made alterations to suit himself. His heads carry his style; the various bodies can be very different. The brant and the round-bottom shelldrakes are totally his, I think, and are excellent in every respect; bold, smooth, and carefully painted in a clean, primitive fashion with no attempt to blend. The bodies are of white cedar, the heads are pine. The eyes of the brant are simply round-head brass tacks while his other decoys have glass eyes.

The small red-breasted merganser drake (Plate 74) in whose head the touch of Richardson is evident seems almost to be a decoy from another area—from Maine or Nova Scotia perhaps. It could be that Richardson made it there, before any Vineyard influence, yet its last paint coat copies Smith as much as the stubby body permits.

73. FRANK RICHARDSON. The hen shelldrake of Plate 72, showing the serrated bill.

Except in the case of his brant, all of which are closely similar, future identification of Richardson's work will not be easy. He was an excellent carver but, given the bewildering variety of improvisation in his ducks, positive identification may rest on such subtleties as the type of weight used, the method of hollowing, the cutting of bill details and, most subtle of all, the "look" of the head.

With Frank Richardson we reach the last of the known older carvers who may be linked however tentatively with the style which began with Benjamin D. Smith or earlier. It was Smith to Chadwick; Chadwick to Pease and Dugan; Richardson with demonstrable ties to Smith.

74. FRANK RICHARDSON. Red-breasted merganser drake. The small body is solid, and Richardson followed Ben Smith's paint pattern very closely (see Plate 84).

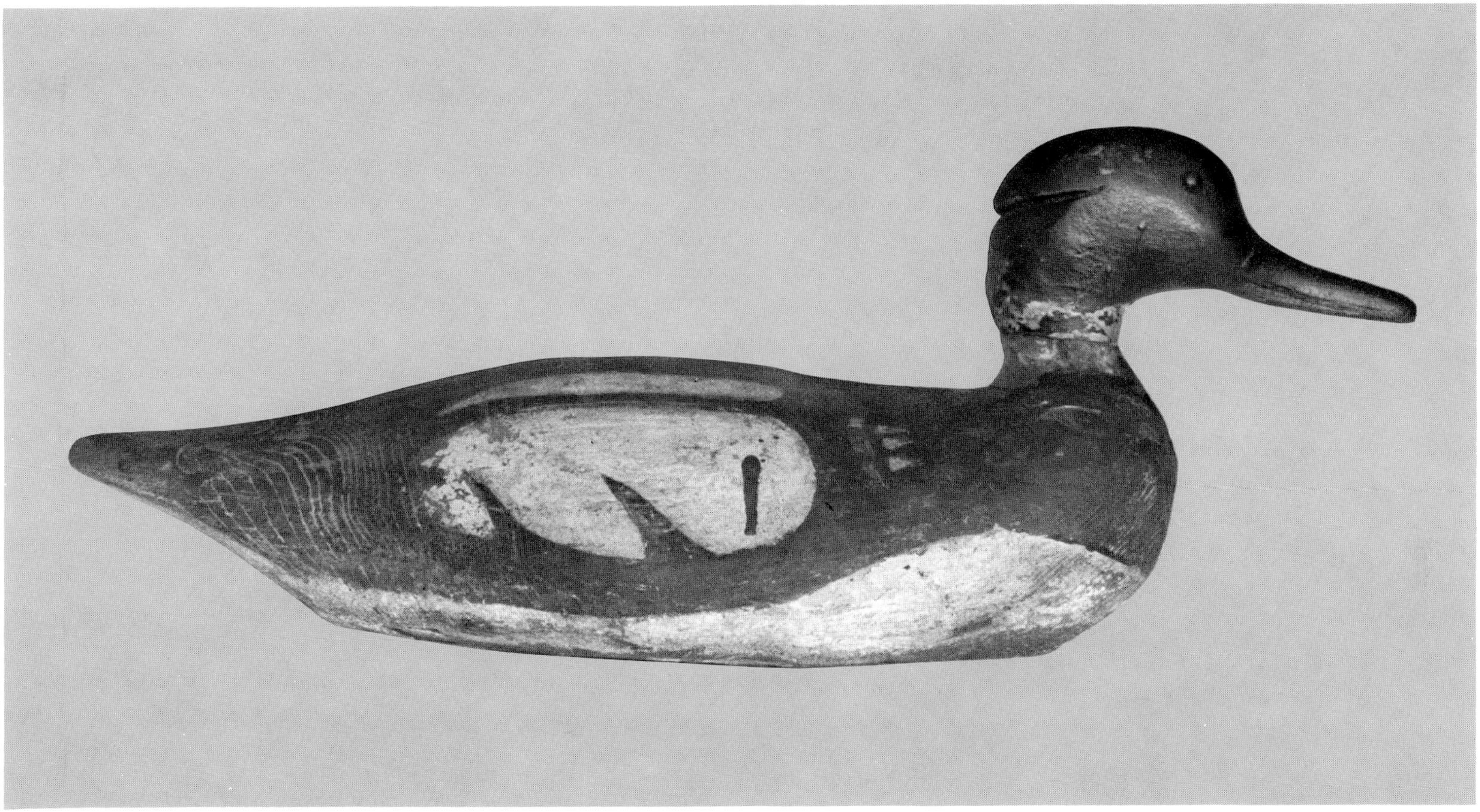

75. FRANK RICHARDSON. Red-breasted merganser drake. Hollowed from bottom. Original paint.

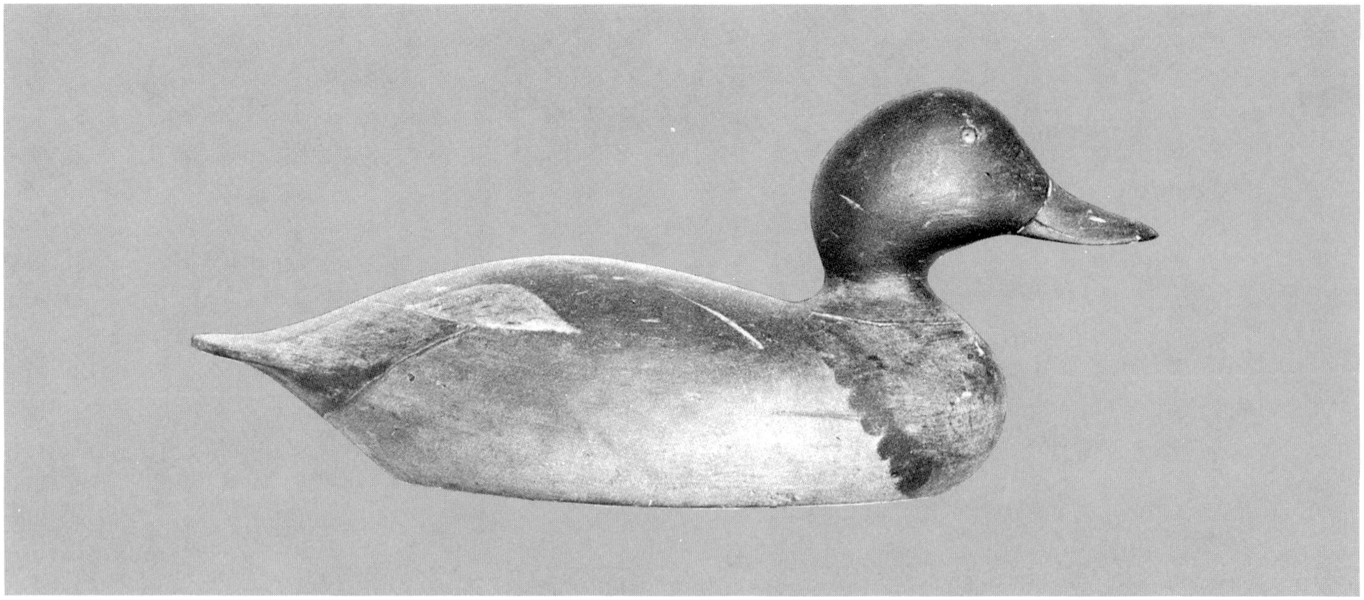

76. FRANK RICHARDSON. *Bluebill drake. A large, hollow bird with a paint pattern (original) that suggests a Mason factory decoy.*

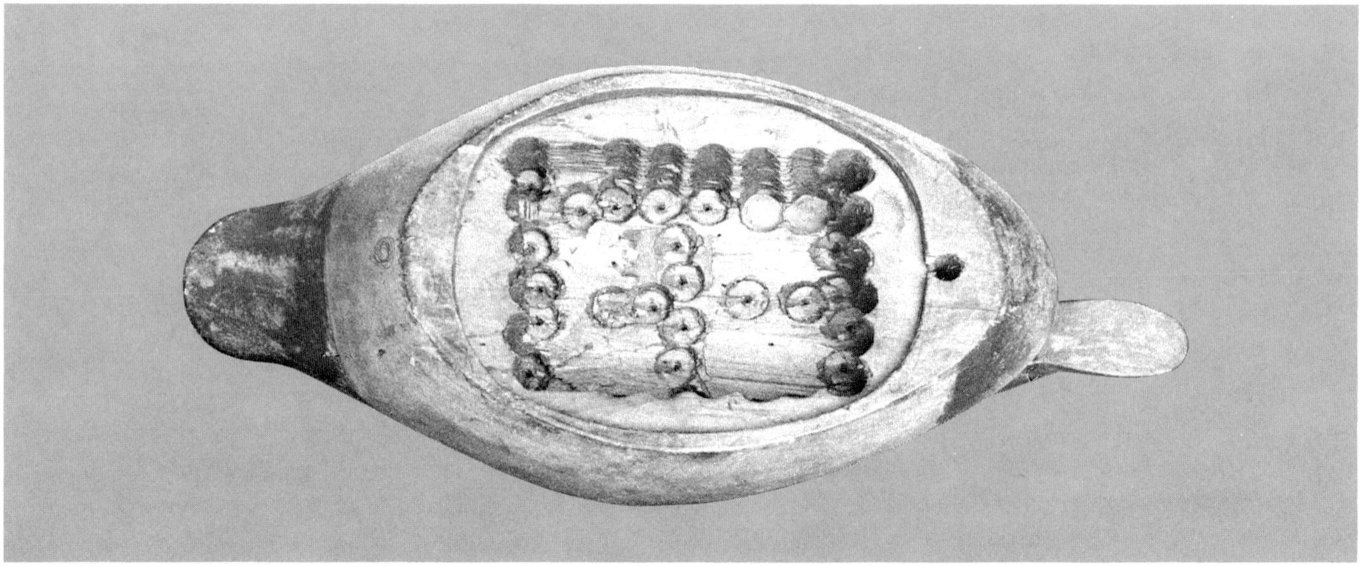

77. FRANK RICHARDSON. *Base plate removed to show the method of hollowing used for the bluebill of Plate 76.*

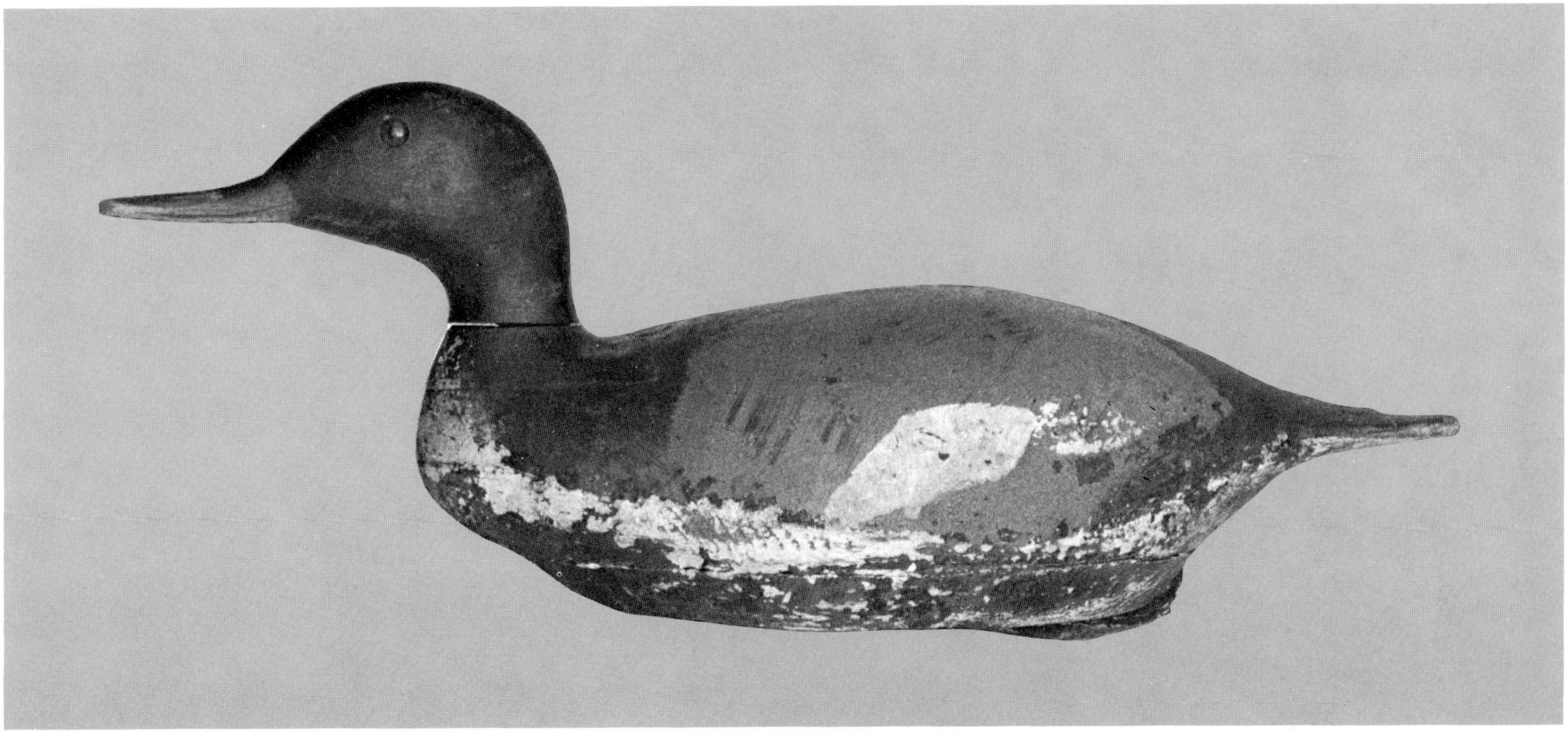

78. FRANK RICHARDSON. Red-breasted merganser hen. The block was sawed apart an inch from the base, hollowed, and screwed together. Original paint.

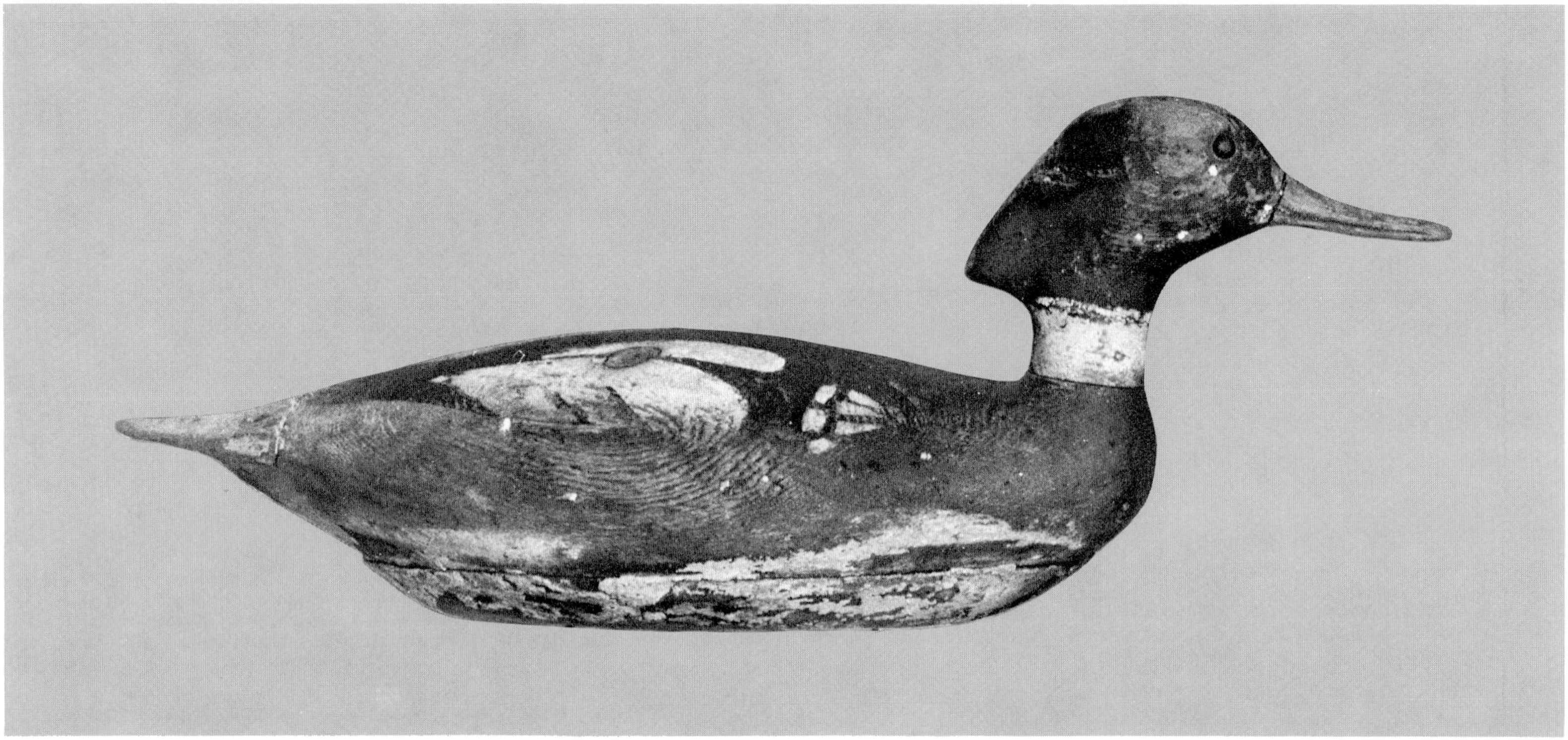

79. BENJAMIN D. SMITH. Male red-breasted merganser. This hard-used beauty, still bearing vestiges of original paint, was at some time owned by Smith's friend, Frank Richardson, who cut his typical, tiny tooth serrations into a replacement bill and hollowed the decoy from the bottom.

80. BENJAMIN D. SMITH. The shelldrake of Plate 79.

Benjamin D. Smith 1866–1946

The Vineyard Gazette of August 13, 1946, carried the following obituary:

"Benjamin D. Smith, retired carpenter, of Oak Bluffs, died at his home on Friday following a prolonged period of failing health. He was eighty years of age and had made his home in Oak Bluffs for many years. He was born in Provincetown, the son of Mr. and Mrs. Obediah Smith. Little is known of Mr. Smith, who was retiring in his habits and lived alone. He was employed as a carpenter during his active years and was famed as a hunter, but talked little about himself or his past. He is not known to have any living relatives.

"Interment was in Oak Grove Cemetery, Oak Bluffs." It is not noted that he happened to carve some of the finest decoy ducks ever seen.

Most of the great names of American decoy carving were professionals each of whose lifetime production amounted to hundreds and even thousands of examples. Their work was appreciated in their own time and they had many incentives to improve technique and refine style over a long period. Smith, on the other hand, was a loner who made his decoys for the ducks only, not for men, yet his carvings are a distillation of great natural talent, the keenest powers of observation, and superb technique.

Each of his birds, glowing with life, has been caught in a logical and transient posture of the duck it represents. Where the typical Chadwick decoy is a quiet thing, solid, imperturbable, serene, the Smith counterpart has a lively alertness; not a quality of alarm (always to be avoided in a bird lure) but a contained wakefulness. The shelldrake is so rakishly shelldrakelike; the black duck so thoroughly a reclusive, wary black duck and no other: the extraordinary likeness of the canvasback is ready to stretch that strong neck and take wing at any sudden movement of the hidden wildfowler.

In season, Ben Smith was a market gunner and appears to have been his own boss, working as a carpenter when he needed money but refusing to be tied down to steady employment. He was noted for

81. BENJAMIN D. SMITH. Pair of hollow canvasbacks (hen above, drake below). Collection of the Duke's County Historical Society.

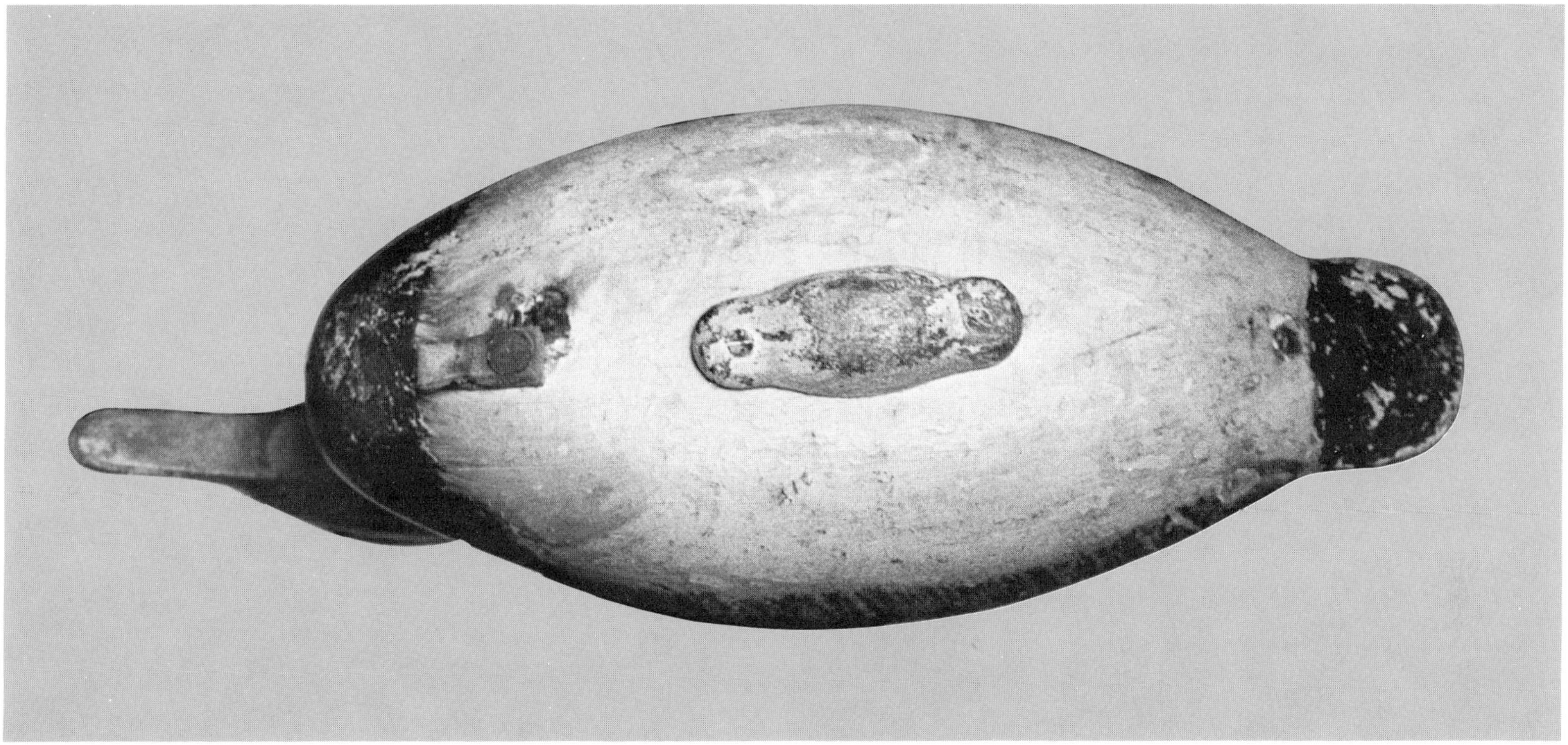

82. BENJAMIN D. SMITH. View of the bottom of the hen canvasback of Plate 81, showing Smith's unique cast-lead weight.

keeping a fine orchard of which he was very proud. An Oak Bluffs man today, a man in his fifties, recalls an incident of his childhood: With several other boys he went one night to steal some of Smith's apples, not realizing that there had been a lot of recent poaching on those trees. Ben was waiting for them in the crotch of one tree with a shotgun and fired rock salt at the boys as they disappeared over his fence. My informant ran in the opposite direction and was almost decapitated by a rope strung across the path for just that purpose. He says it was the last of the apple thievery.

Everyone I have been able to find who has any memory of Smith agrees on one essential point about him: He was not a man to trifle with. He was known to make spot judgments of people on brief acquaintance and, as one man reports, "If he decided for some reason, any reason, to dislike you, that was *it.* He would have nothing further to do with you."

He was tall, lean and strong and quick in movement, a small-eyed man with a pointy nose and heavy moustache, a man who habitually wore a felt hat. Briefly, obscurely, and late in life, he married. It quickly, and probably mercifully, ended in divorce. Sooner or later he quarrelled with everyone in his orbit

83. BENJAMIN D. SMITH. The drake of the hollow canvasback pair in Plate 81. The block was separated for hollowing on a diagonal line which descends from the upper back to just below the tail.

and one source of information reports that he ended up with a grudge against even Keyes Chadwick though the reason, if one did exist, is lost in antiquity.

There could be no better evidence of Smith's capacity for vindictiveness than his last will and testament. It is a one-page document in three parts; part three, with all of its punctuation, capitals, and underlining as it appears in the original, but with the name and address of the brother omitted, reads as follows:

"I give and bequeath to my brother (?) who I publicly declare without fear of successful contradiction to be the most contemptible thief it has been my misfortune to meet, the sum of $2500. The above mentioned sum being the amount he cheated me of to feather his own nest in his capacity as Executor of the estate of my sister Mrs. Mamie E. Talbot who at the time of her death was a resident of Provincetown, Mass.

84. BENJAMIN D. SMITH. Male red-breasted merganser. Rounded bottom, solid body, original paint.

"The above mentioned still having in his possession the above mentioned $2500, I take this method to release him, if possible, of the stigma of being a contemptible Thief."

There, brethren, is one way of settling accounts. The executor of this will recalls that he had to promise that, upon Ben's death, he would see to it that the will was published in the brother's hometown newspaper. When the time came, the executor failed to keep his promise. "His brother had become a man of some parts by then, and had his own family," he explains, "and I didn't have the heart to do what Ben asked. What good could have come from such a thing?" What good indeed.

Another inclusion of that will and testament reveals a facet of Ben Smith that now defies all further amplification. Noone recalls that he was a particularly musical man, but the will, part two, states:

"I give and bequeath to my cousin Mrs. Geo. Burke (nee Edith C. Watt) all right and title to the copyright song entitled 'A Song of Sonja Henie, the Golden Girl', same being recorded in the Library of Congress at Washington, D.C."

Smith's decoy bodies are of white cedar, the heads of sugar pine. The Black Duck is flat-bottomed but most of his other species that I have seen vary from rounded bottom to flat. Some of the bodies

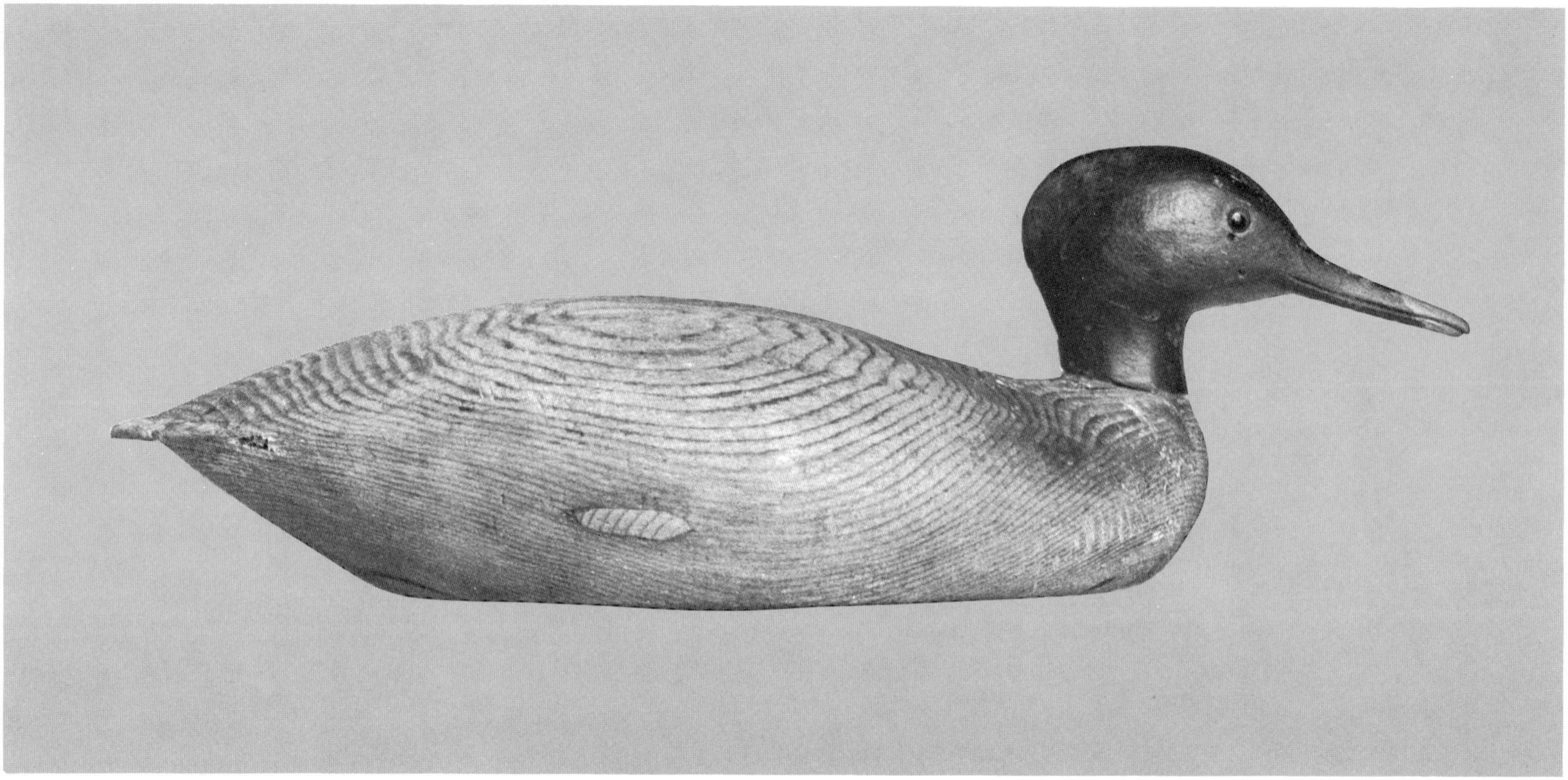

85. BENJAMIN D. SMITH. Female red-breasted merganser. The head bears the original paint but the badly worn body paint has been removed. The body repair plug is typical of Smith.

are solid, others are hollow. There are examples with metal-stud eyes, painted eyes, and glass eyes, perfectly set. His lead weight is unique in shape and, when present, is an instant identifying feature (Plate 82). He doesn't appear to have used it on his flat-bottomed decoys however.

The pair of canvasbacks in the collection of the Dukes County Historical Society in Edgartown is proof that Smith was an excellent painter, too, subtle and accurate. They are the last decoys he possessed and were purchased from him in his final year by an Edgartown collector who found him, that day, up in one of his apple trees doing a little pruning.

When he went gunning, generally on the great tidal pond known as Sengekontacket (and by some as Anthier's) between Oak Bluffs and Edgartown, he is remembered as wrapping each decoy separately in straw for the wagon trip. This indicates a rather small rig, and concern for his fine blocks. The average hunter bundled his decoys loosely into grain bags where one banged against the other, with broken necks, tails, and anchor scars the inevitable result.

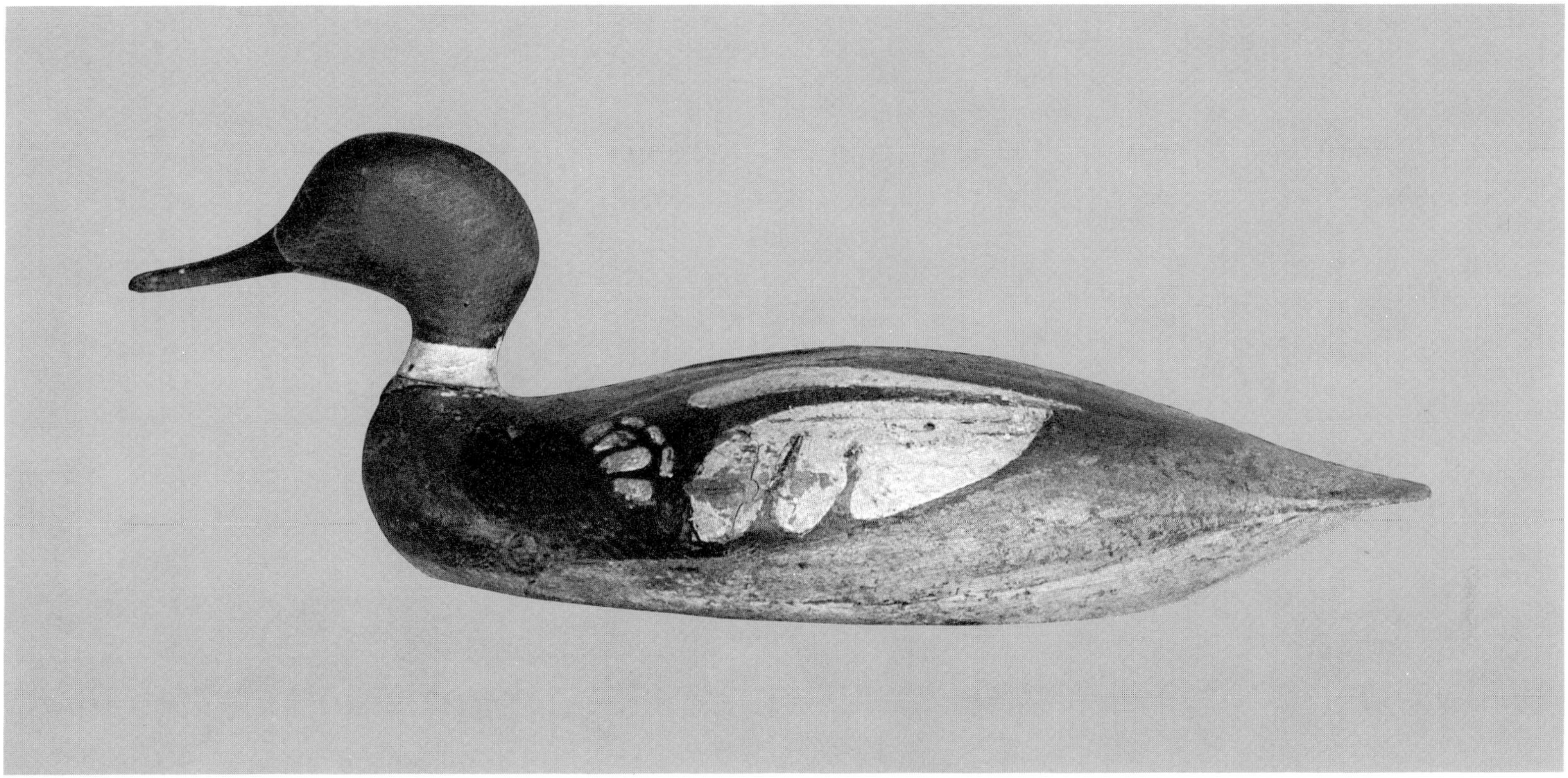

86. BENJAMIN D. SMITH. Male red-breasted merganser (shelldrake). Solid body, flat bottom, original paint. This body style was the prototype for much of the work of Keyes Chadwick.

The decoys in the accompanying photographs include, with three exceptions, all of the examples of Ben Smith's work that I have seen. A market gunner, and Smith was one, needed a fair number of decoys. Where are the Smith redheads and bluebills, the most common ducks of his time? Did he carve any brant? Or goldeneyes, widgeon, or Canada geese? Perhaps, like the wildfowl they imitated, these wooden birds migrated too. In the 1930's and 40's second-hand dealers made frequent visits to the Island and filled their pick-up trucks with old decoys purchased at 25 cents apiece.

Time and, hopefully, the gradual appearance and identification of more of his work will establish the reputation of this very gifted man. Let all who would pursue the enjoyment of the decoy as Art consider these words about Ben Smith spoken by his contemporary and one-time apprentice Keyes Chadwick: "He taught me my craft, and there was literally nothing Ben couldn't do very well. He was the *best*."

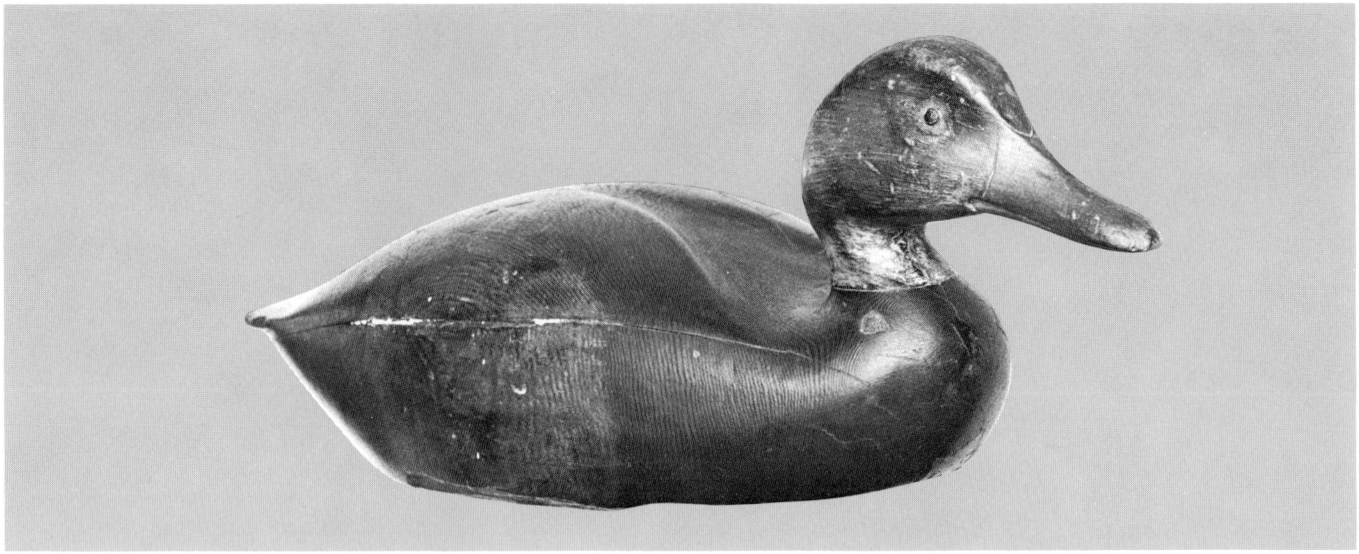

87. BENJAMIN D. SMITH. Black duck. The eyes are metal studs. The head was carefully painted but the body was merely stained a deep burnt-cork brown. There is no sign that the block ever was weighted. Bruised, cracked, and gouged over the years, it remains an extraordinary work of art.

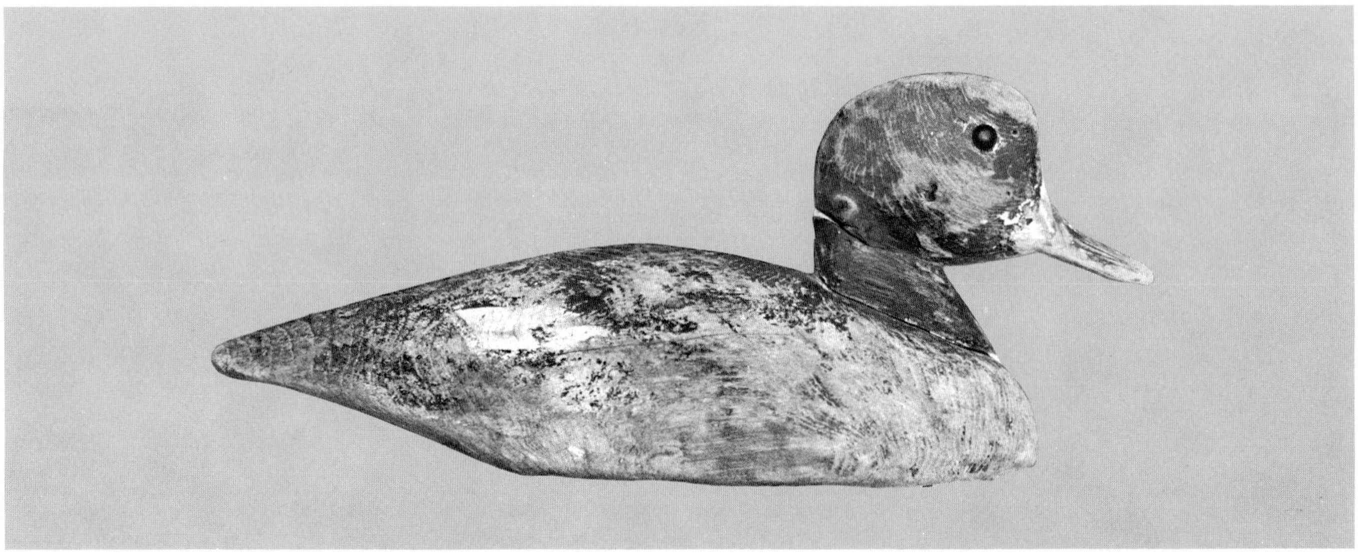

88. EDWIN T. SMITH. Bluebill hen. A hard-used little bird that saw many a cold sunrise on Katama Bay.

Edwin T. Smith 1878–1963

Edwin Smith (no relation to Benjamin D. Smith of Oak Bluffs), an Edgartown man, worked the old family farm on Katama Bay and was seasonally a commercial fisherman and market gunner. His son still has Smith's ancient ten guage damascus-barrelled Parker and his rig of decoys, both ducks and shorebirds. The duck decoys are a mixed bag of varieties and styles but here and there is an example of Edwin Smith's carving.

89. ALTON TUCKERMAN. American merganser drake. A large, bold, hollow carving. The stark black and white paint coat has been removed but the pattern of it remains. The bottom plate was sealed to the base with a gasket of thick white lead before nailing.

Alton C. Tuckerman 1877–1968

Born in New Bedford, Tuckerman came to Vineyard Haven as a boy and learned the tinsmith trade which he practiced until the last decade or so of a long life.

He is represented in the world of carving by only three decoys that I know of, all of them big, hollow, wonderfully strong and primitive likenesses of the American merganser, two males and one female. Some time after the discovery of these three decoys I had a conversation with Tuckerman who said that at one point he'd owned many more ducks of his own manufacture. But, as he put it, "When they put the law on the birds I just left 'em in the blind and never went back." He was disgusted with the federal government for presuming to tell him how many ducks he could kill, so he stopped wildfowling then and there. And the time was not in recent generations when low limits (and prohibition against any killing of certain species) are the rule, but back in 1918 when restrictions were initiated; when a man could no longer shoot all the ducks he chose to shoot, but was allowed only a miserly twenty-five a day.

Ah well. Let us hope that the Tuckerman decoys fell into good hands, that they didn't slowly decay into the marsh, and that we may see some of them by and by.

The bodies are of white cedar hollowed from the bottom which is covered with a half-inch plate of wood to seal out the water. Heads are pine and the whole appearance is one of bold ruggedness. Marks of the drawknife and pocket knife are everywhere still present . . . no attempt at complete smoothness was made. The paint on the specimen illustrated here was removed long ago but the outlines of the separate strongly defined areas remain clear.

It may be wondered why the mergansers are represented on Martha's Vineyard in such comparative abundance. Their quality as a table bird is best described, I think, in the words of Forbush, the ornithologist—and onetime hunter. Writing in 1912 he said of the American merganser:

"This bird, when cooked in the ordinary way, is about as palatable as a stewed kerosene lamp-

90. ALTON TUCKERMAN. A detailed portrait of the American merganser male in Plate 89.

wick, but some people on the coast are able to prepare and eat a shelldrake now and then with a clear conscience. There are some hardy gunners and fishermen whose appetites are so good that it is immaterial whether they are eating flesh, fish or finnan haddie, and I have been credibly informed by some of these enthusiastic coast gunners that they actually enjoy eating a shelldrake or two in the spring after a hard winter."

I have put the question to a few old hunters. The answer of one of them was simply that "I love to see them sweep in to the decoys the way they do, so low and rakish." Another had what I believe to be the logical explanation. He pointed out that in late winter (back when the open season was considerably longer than it is now) when even the salt ponds iced over, there were usually airholes left open at the upper end of the coves. When the other duck species had been pushed farther south, the mergansers, fish-eaters entirely, could still find their food in those airholes which often reached to the shore where the camouflaged hunter waited. Even when the airhole was far out in a pond, a gunner dressed in white could sit quietly on the ice edge and get excellent shooting.

91. LITTLETON C. WIMPENNY. Goldeneye drake. Duke's County Historical Society Collection.

Littleton C. Wimpenny 1845–1916

Wimpenny, an Edgartown man, is best remembered as the Registrar of Deeds of Dukes County, an office he held for the last 25 years of his life. As a younger man he went to sea with his master-mariner father and served as purser on a steamship.

There is no mention of his hunting or carving activities in any account that I can find but it is certain that he engaged in both. Years ago I was informed by Keyes Chadwick that Wimpenny had made duck decoys for the sporting goods house of Wm. Reed & Son in Boston. They were carved from broken cedar spars which were obtained from New Bedford shipyards and they brought Wimpenny $9 per dozen, painted. Chadwick also said that Wimpenny was the first man he ever saw make a decoy.

The decoy illustrated here is the only example of his style that I have seen but many others must survive. This cock whistler, almost all of the paint worn from the head, is a solid, unimaginative, sturdy work with no delicate or projecting details of carving to be broken off during a life of hard use. One wonders if possibly it is patterned after a factory decoy.

Wimpenny's signature on his Will is a noble example of fine calligraphy. The sure, graceful touch of his hand suggests that carving in wood must have come easily to him.

The Unknown Carvers

92. UNKNOWN CARVER 1. I believe this merganser to be the oldest Vineyard duck decoy I have seen. It is hollow, studded with shot, has a wrought-iron nail or two still in her bottom, and was covered with about ten coats of paint, all bad, when found.

The Unknown Carvers

As might be expected, the number of unknown carvers is greater than the known. Further, I have reason to believe that if time and permission allowed the search of various other cellars, barns, attics, garages and fish-shacks, the number could be more than doubled.

I have yet to see an old gunning rig that doesn't include a fair portion of strays—decoys found on the beach or in the marsh during a shoot, or later. Decoy anchor lines broke (no nylon in those days) and were subject to rot and would weaken greatly after a few seasons of use. When one or two drifted away across wide ponds or bays during heavy weather (always welcome to a wildfowler), they were seldom pursued because the time could be better spent shooting more ducks. They were more valuable than decoys.

In most cases the unknown styles are as distinct and individual as men's thumb prints, but there are some which strongly suggest the works of known makers, and may even be their early attempts or experimental types. A case in point is the trio of bold, strong birds carrying the RENEAR brand (Plate 102). They magnify the stylistic traits of Jim Look but have subtle differences. Walter Renear was a noted Vineyard Haven gunner, a very quick and accurate shot, and a close friend of Jim Look, but he is not known to have done any carving and so these excellent fellows must be relegated to the limbo of Unknown Carvers.

The following is a list of men who are known to have made decoys on Martha's Vineyard but whose work cannot positively be linked to their names at this time:

1. Frank Goodwin, Gay Head Indian who lived in Oak Bluffs.
2. Charles A. Joy, Oak Bluffs.
3. Sam Kidder
4. William King, Edgartown
5. John L. Mayhew, Oak Bluffs.

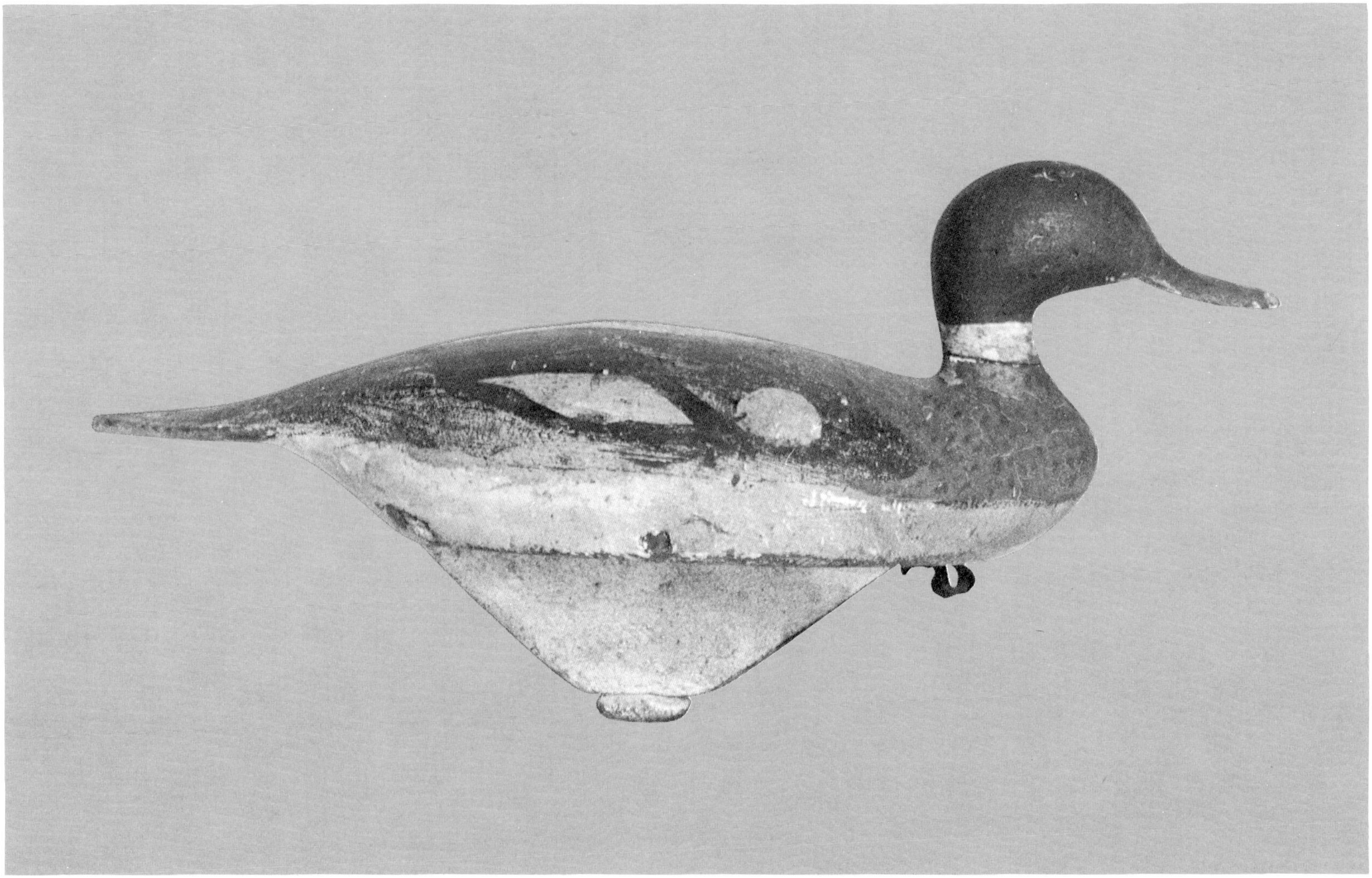

93. UNKNOWN CARVER 2. The Long-Tail Carver. A male red-breasted merganser. The keel was added by a later hand, an alteration common to Edgartown gunners.

6. William Pent, Edgartown
7. Lyman Spencer Smith, Edgartown
8. Tip Smith, Oak Bluffs.
9. Arthur C. Vincent, Edgartown

Years ago I saw a set of geese made by Charles Joy (1851-1935). They were very crude and were heavily branded CHAS. JOY on the bottom.

94. UNKNOWN CARVER 2. The Long-Tail Carver. Bluebill male. Note the fine cutting of detail in the head. This man was exceptionally good. Paint is original; the keel was added by a later hand. Collection of the Duke's County Historical Society.

95. UNKNOWN CARVER 2. The Long-Tail Carver. A red-breasted merganser (above) in original paint, and an over-painted cock whistler. Bodies are broad and have excellent stability in rough water. Weights are flat rectangles of lead nailed well to the rear of midpoint.

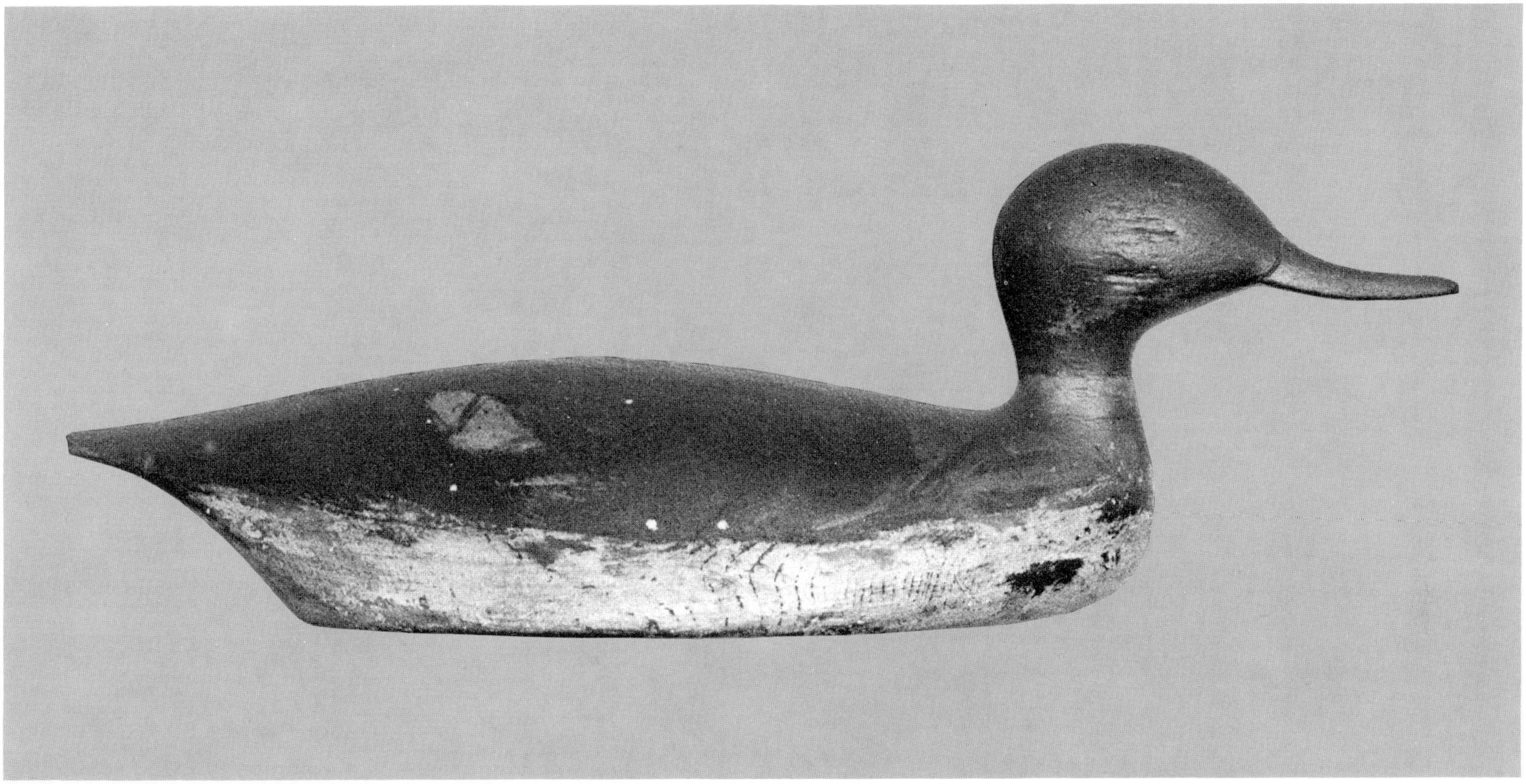

96. UNKNOWN CARVER 3. Red-breasted merganser hen. I have seen a number of these attractive shelldrakes and strongly suspect that they are the work of Frank Adams, but there are stylistic differences from his other ducks — painted eyes, rounded breast, curved bill — that make definite attribution impossible at this time.

97. UNKNOWN CARVER 3. An American merganser drake head mounted as a paperweight.

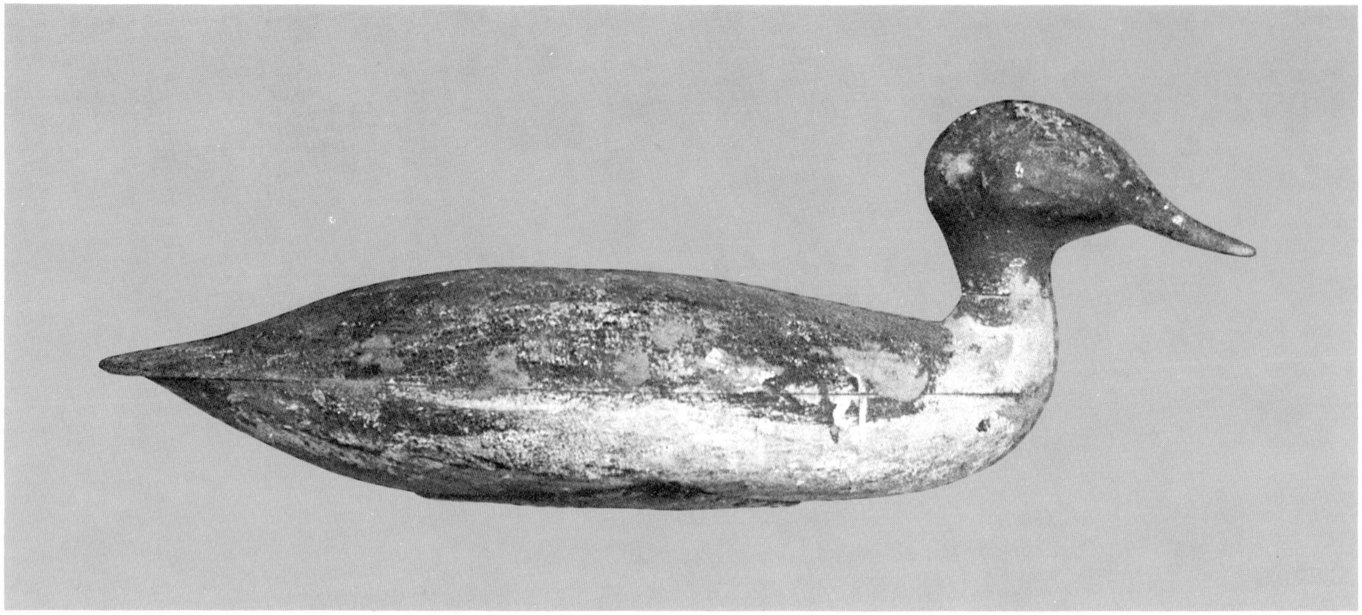

98. UNKNOWN CARVER 4. American merganser male. A fine, hollow decoy in very worn original paint.

99. UNKNOWN CARVER 4. The American merganser of Plate 98, showing the unique, six-sided weight which also appears on the ruddy duck of Plate 100.

100. UNKNOWN CARVER 4. While crudely overpainted as a bufflehead, this bird was made, beautifully made, for a ruddy duck. The original blue of the bill, the red-brown of the body paint, and the notched tail base identify this hollow rarity. I believe it to be of Chilmark origin, one of the few places on Martha's Vineyard where this species appeared in quantity.

101. UNKNOWN CARVER 5. Redhead female (bottom). For size comparison this enormous lure has a normal-sized Canada goose (Plate 66) on her back. Heavily weighted, and chained to an anchor, these crude giants were placed in the center of a pond and left out until the season ended. The idea was to catch the attention of a far-off flock and sucker them close enough to see the more natural decoys inshore. A mixed stool of redheads and bluebills was used.

102. UNKNOWN CARVER 6. James Look style. Black ducks. While closely resembling Look's work (Plates 58-60), these heavier decoys are by another hand and must be regarded as Look prototypes or copies. All are branded RENEAR, and Walter Renear of Vineyard Haven, while not known to have carved decoys, was a hunter and close friend of Jim Look.

103. UNKNOWN CARVER 6. James Look style. Another view of the middle bird in Plate 102.

104. UNKNOWN CARVER 7. The man who carved this odd and wonderful and tiny red-breasted merganser female was an ex-slave who worked for a retired sea captain, Benjamin Cromwell of Vineyard Haven, for some years after the Civil War.

105. UNKNOWN CARVER 7. A detail of the little merganser hen in Plate 104 which clearly shows the mortising of the neck to the body.

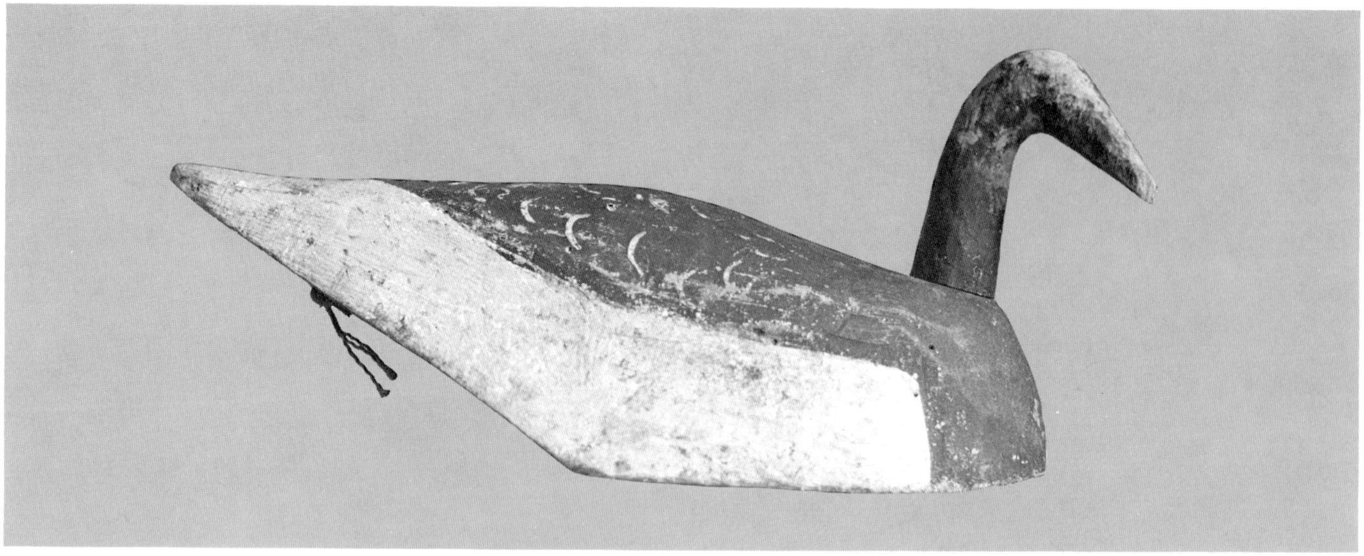

106. UNKNOWN CARVER 8. A brant of marvelous strength and simplicity whose body appears to have been carved with no other tool than a hand-axe. This individual was anchored from the aft end to give further realism to the stool.

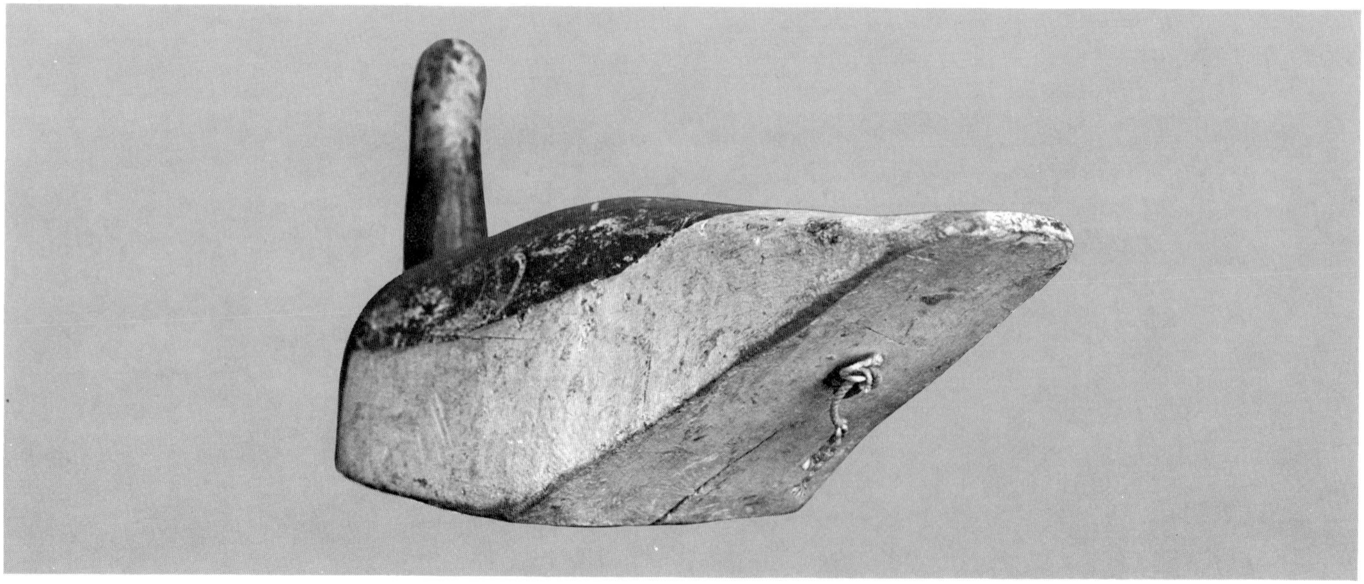

107. UNKNOWN CARVER 8. The brant of Plate 106. It bears the brand of Levi Jackson, a famous Edgartown fisherman who, however, made no decoys.

108. UNKNOWN CARVER 9. A monumental redhead drake of judicial dignity. Considerably larger than life-size.

109. UNKNOWN CARVER 9. The redhead of Plate 108. The bill is one and a half inches wide at the base.

110. UNKNOWN CARVER 10. Bufflehead hen, much overpainted. Mindful of the work of Ben Pease (Plates 67-69). And he lived in rich bufflehead territory.

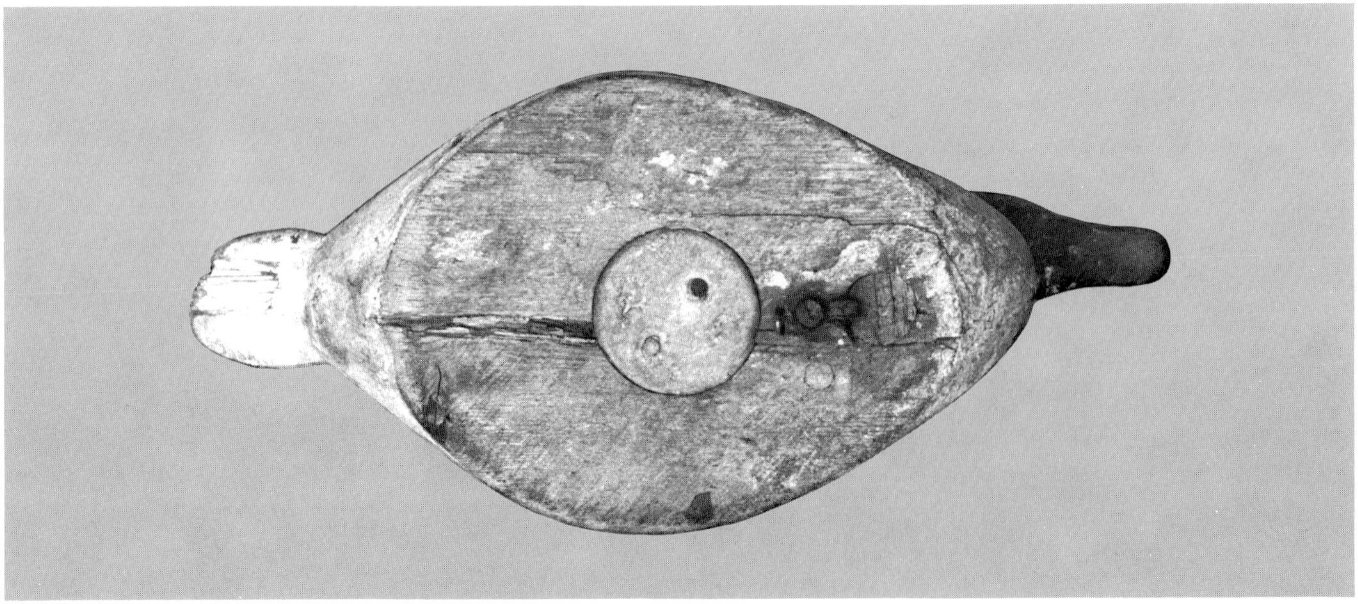

111. UNKNOWN CARVER 10. The bottom of the Plate 110 bufflehead.

112. UNKNOWN CARVER 11. Black ducks. This fine, hollow pair may be of Edgartown origin. Duke's County Historical Society Collection.

113. UNKNOWN CARVER 12. Black ducks by a master craftsman. Late Nineteenth Century. Original paint.

114. UNKNOWN CARVER 13. Bufflehead overpainted as a goldeneye drake. This diminutive bird is almost certainly from Edgartown, the one Island area where bufflehead were regularly hunted.

115. UNKNOWN CARVER 13. Front view of the bird in Plate 114.

116. UNKNOWN CARVER 14. An ancient hen red-breasted merganser. The two-piece body is joined with wrought-iron nails.

117. UNKNOWN CARVER 14. A detailed look at the female red-breasted merganser in Plate 116.

118. UNKNOWN CARVER 15. American merganser female (above) and male. The bodies are as long as those of Canada geese; the bottoms are round and the decoys must have rolled badly in a sea.

119. UNKNOWN CARVER 16. Mergansers, badly overpainted.

120. UNKNOWN CARVER 17. Probably a redhead. Wings are carved in relief. Part of the neck is a recent repair.

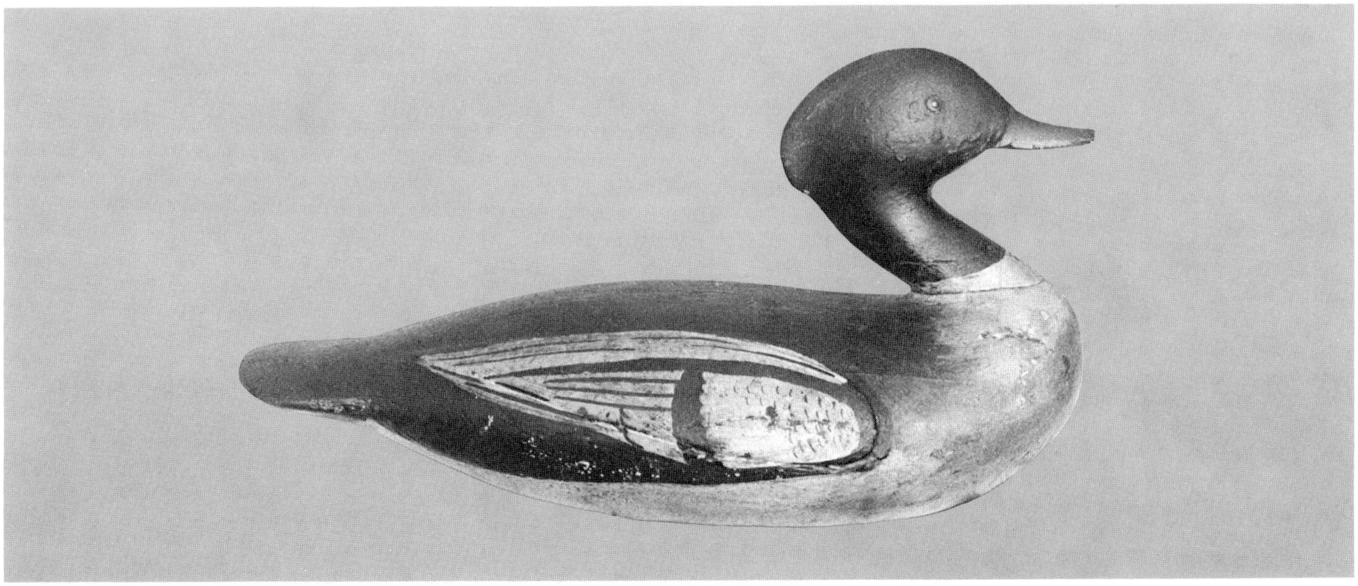

121. UNKNOWN CARVER 18. Red-breasted merganser drake. This gaudy fellow should be seen in color; the breast is a sunrise. Wings are relief-carved.

122. UNKNOWN CARVER 19. Black duck. There is something majestic to me about this big, heavy bird with all its draw-knife and adze marks. The artist must have been a boat-builder; the bow-end is that of a catboat. Collection of Duke's County Historical Society.

123. UNKNOWN CARVER 20. A rather plain little duck of uncertain species, but sound as a nut.

124. UNKNOWNCARVER 21. Baldpate drake (foreground) and hen. Bodies are made of two solid pieces screwed together.

125. UNKNOWN CARVER 22. Probably a redhead female. Badly worn original paint.

126. UNKNOWN CARVER 23. Bluebill hen.

127. UNKNOWN CARVER 24. This is a very old decoy with cut-nails used to spike head to body. Paint is original, and the colors and pattern conform to what would be a great rarity — a Labrador duck in juvenile plumage.

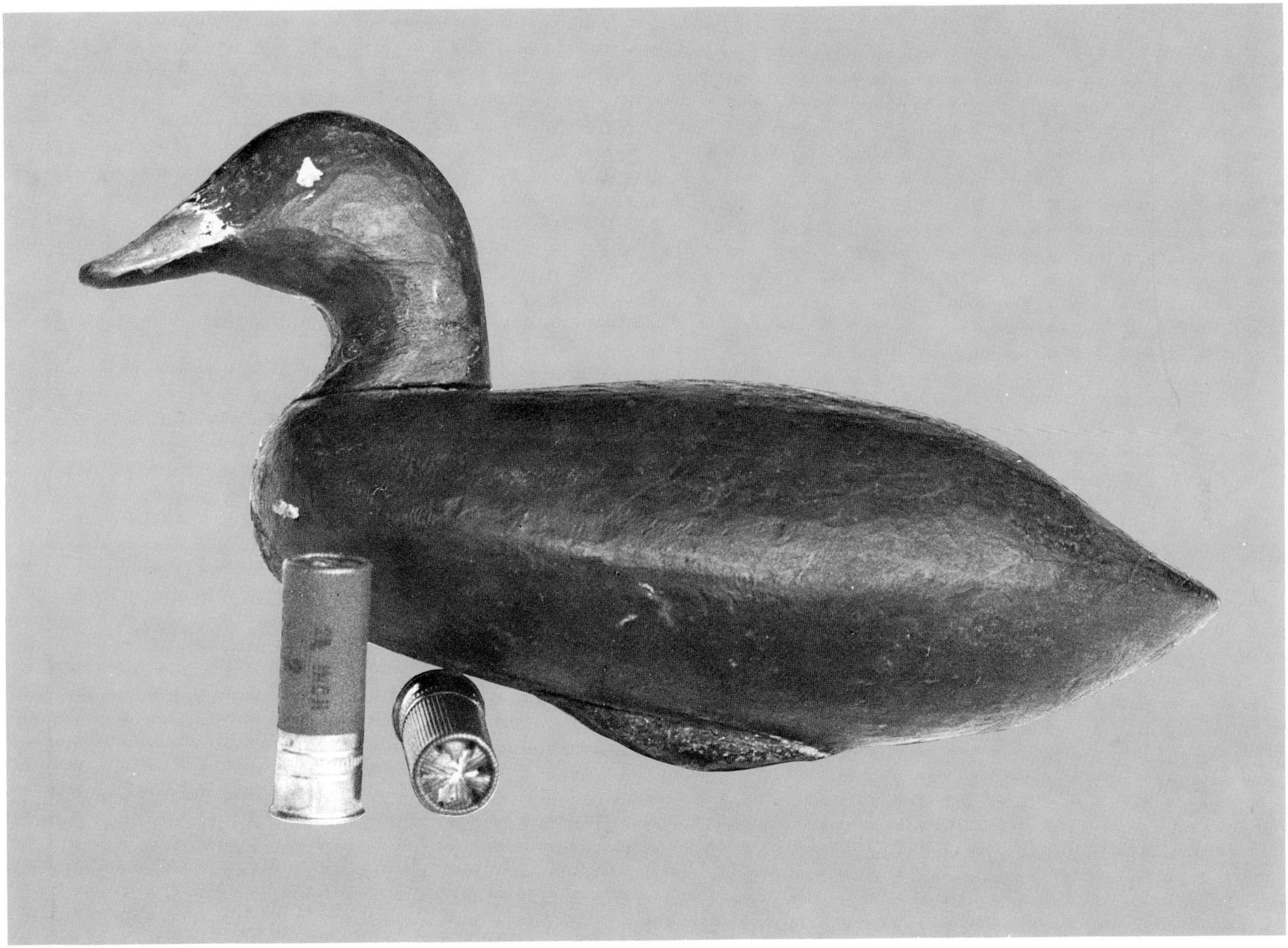

128. UNKNOWN CARVER 25. A tiny duck of undetermined species.

129. UNKNOWN CARVER 26.

130. UNKNOWN CARVER 27. Bluebill male (top) and female. These are said to be the work of Benjamin Cromwell III of Vineyard Haven but I doubt it.

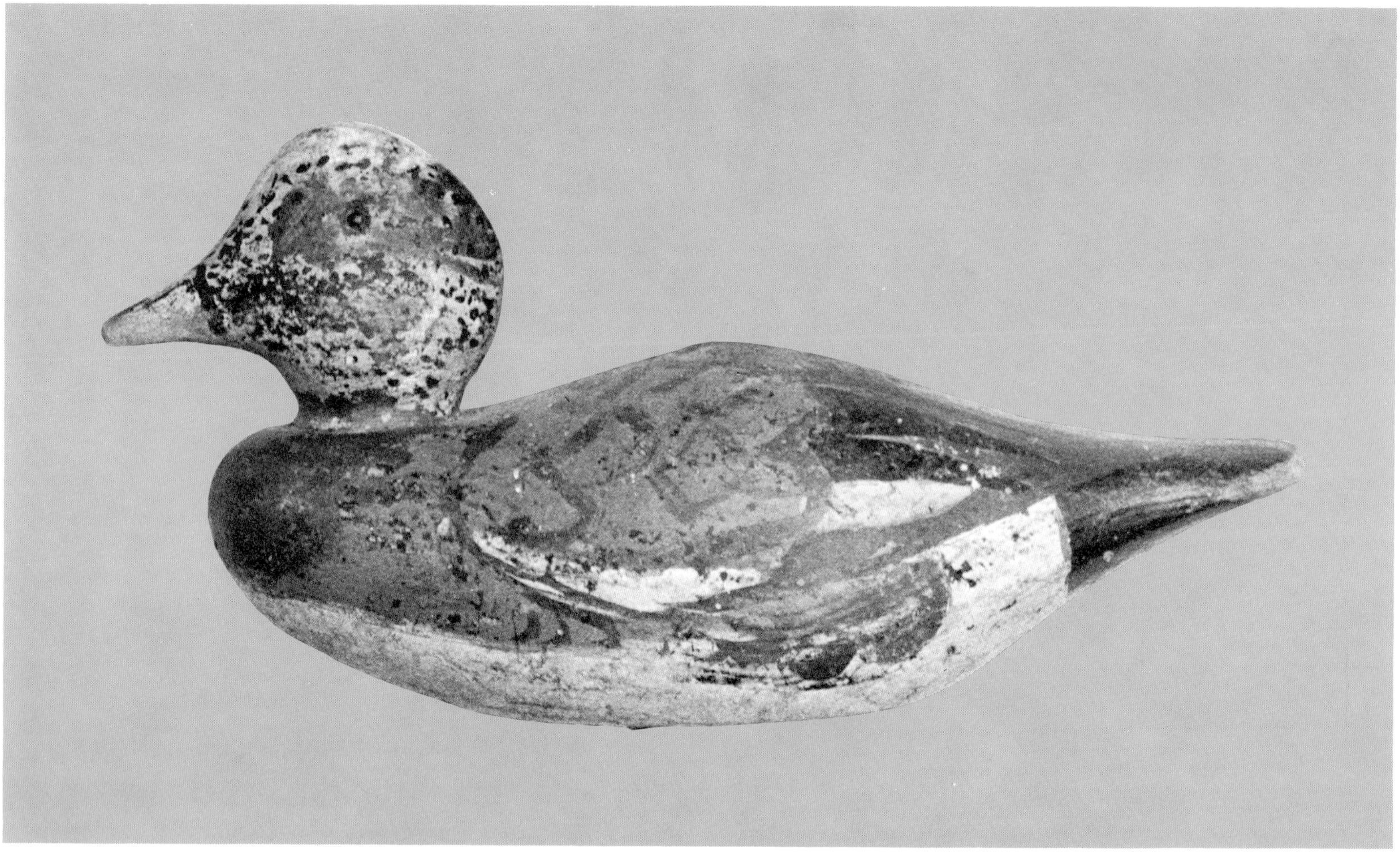

131. UNKNOWN CARVER 28. A poorly overpainted very small, solid decoy, originally a bufflehead or teal.

132. UNKNOWN CARVER 29. Canvasback drake. This excellent carver knew his bird well. Could be the early work of W.E.Dugan (page 61), unweighted, unbranded.

133. Brant. In 1920-1930 over a hundred of these brant were made for use on Martha's Vineyard. Working from a pattern drawn by a member of the Hart family, men in the family tool factory in New Britain, Connecticut made these decoys for the Harts to use on Anthiers Pond, Oak Bluffs.

134. Where have all the shore-birds gone? A proud young West Tisbury hunter, Charles Foote, with the results of a morning shoot in 1918 or earlier. Over forty greater and lesser yellowlegs.

Shorebirds by Unknown Carvers

The lures illustrated here were all found on the Island and have, with few exceptions, seen hard use, were shot over and, at times, into. It may be that certain of them were not carved by Vineyard men. The group of four plover (Plate 135) certainly suggests the fine hand of Elmer Crowell, particularly the marvelous "feeder". If they are Crowell's, they are early examples of his work because they were in use here well before 1900.

As noted in the captions, some of these decoys are of ancient lineage, possibly older than any of the duck decoys in this book. In one case the man who owned the little birds had shot over them before the turn of the century; years before that his father had used them and, he said, "They were old even then!"

The great flocks described by the old gunners are long gone, and other species than the Eskimo curlew would now be nothing but a memory if a federal law of 1918 had not forbade any further killing of them. They were shot in the spring and the fall and it took just a few nondescript decoys, properly placed, to bring them close to the gun. A hunter who could call well, who could imitate the whistles of the various birds, could entice a flock of, say, thirty yellow-legs to his stool, shoot down six or eight of the closely bunched birds on the first pass, then whistle the remainder back repeatedly until the entire flock was wiped out. Such innocence demanded protection but it came late, too late for some.

135. UNKNOWN CARVER 30, before 1880. A quartet of superbly carved plover, either black-bellied or golden, subtly painted for the Fall season. The eyes are a glossy, resinous substance. All of the bills are old but only that of the second from the left is original. These lovely decoys could be the early work of Elmer Crowell of East Harwich. They were in use on the Vineyard long before the turn of the century.

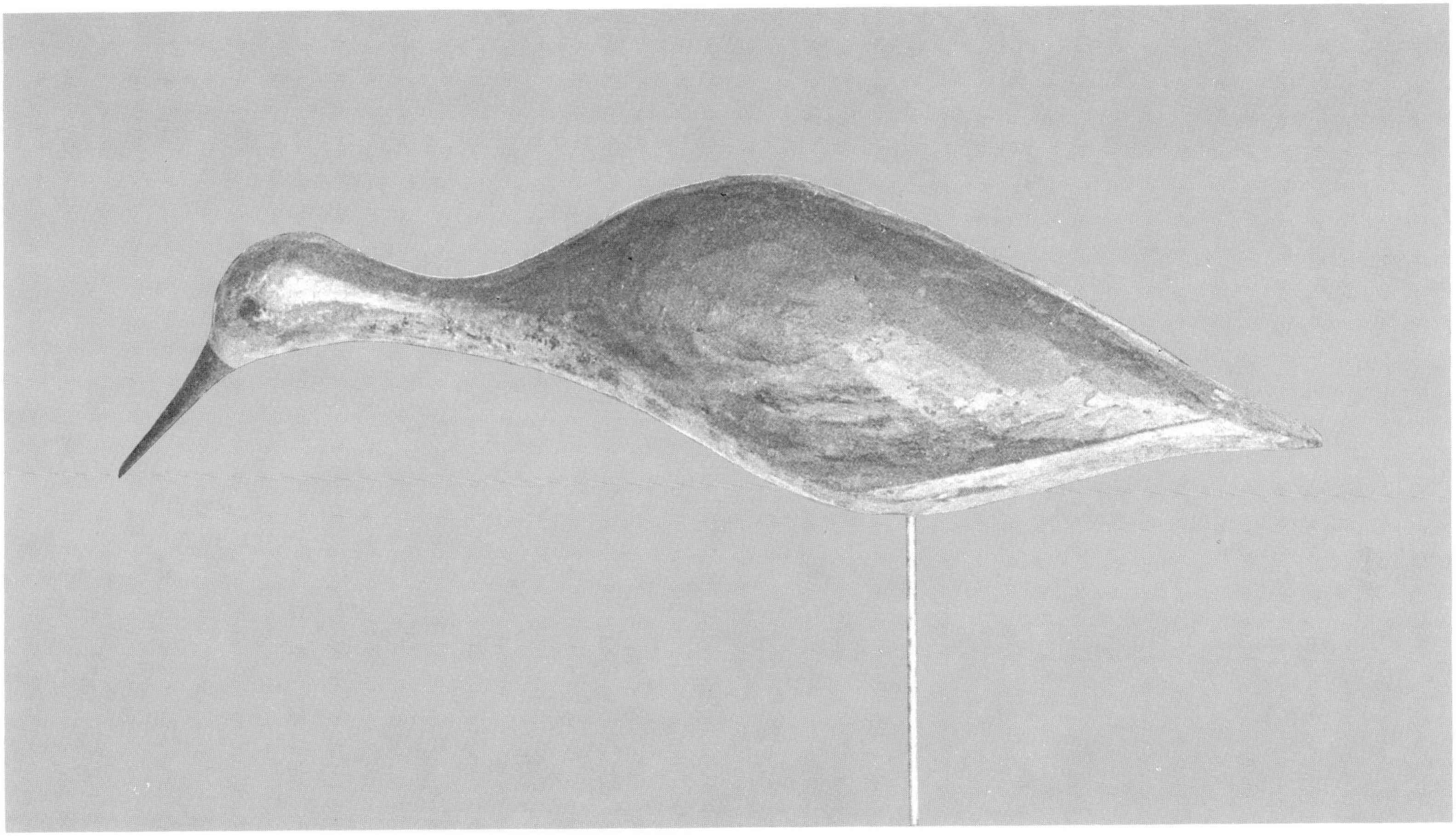

136. UNKNOWN CARVER 31. Greater yellowlegs with wing-tips in relief.

137. UNKNOWN CARVER 32. Based upon a similar plover owned by Winthrop Norton (page 87). I believe this is the work of John L. Mayhew of Oak Bluffs, the one Vineyarder known to have carved working shorebirds for sale. Norton's father learned the tinsmith trade from Mayhew and they were good friends. Still, the attribution must remain uncertain.

138. UNKNOWN CARVER 33. A tiny, finely carved sandpiper. Birds of this size were skinned out, not plucked. One individual provided two mouthfuls; two small mouthfuls. Not worth the cost of a shell unless a number could be slain with one shot.

139. UNKNOWN CARVER 34. An ancient shorebird, species unknown. The wings are cut in relief.

140. UNKNOWN CARVER 35. Black-bellied or golden plover in fall plumage. Gunned over by several generations of Edgartown fowlers.

141. UNKNOWN CARVER 36. This bird is somewhat of a mystery. It is large for a plover and hollow. The tail has been shortened by a later hand; the bill is not original; eyes are pitch. But once upon a time it was a well-made beauty.

142. UNKNOWN CARVER 37. These hard-used, shot-over plover were used on the flats of Edgartown Great Pond by Edwin Smith (page 111). Eyes are brass studs, the fills long-gone.

143. UNKNOWN CARVER 38. More plover that were gunned over by Edwin Smith (page 111) and may have been made by him. The nail bills are original.

Conclusion

The means by which our known carvers have been identified are varied and, in some cases, may be questionable. I have tried to avoid taking certain rather shaky old memories as fact and have refused to credit the "I assume . . .", the "I guess , , ,", and "I think so . . ." without rigorous research. Still, the appearance of a supposed Ben Pease bluebill pattern inscribed with the name "L.W. Wimpenny" in a box of Chadwick patterns gives one pause (Plate 70). Such riddles seem beyond solving.

It was Keyes Chadwick who first identified for me the work of Benjamin D. Smith. Tears welled in his eyes when I showed him the exquisite black duck of Plate 87. The decorative birds of Frank Adams were known to many but it was only his brother-in-law who recognized his working decoys; they had been obtained directly from Adams.

Of men such as Jackson, Hancock, Norton, and Tuckerman there can be no doubt, but what of the variety displayed in the carvings traced to Richardson? In his case I submit the evidence and leave a final verdict to time.

I suspect that the list of Vineyard carvers will grow. Somewhere among the people I missed talking to will be an elderly lady who can produce an old photograph of her father or grandfather standing proudly, gun over the crook of his arm, in front of a barn door whereon about forty ducks are hanging by their bills. Near his feet, along with his retriever dog, will be a pile of decoys which he is known to have carved. Perhaps one will be bottom-up to reveal a large, hexagonal sheet of lead for its weight (Unknown Carver Number 4); or maybe the tails of the wooden birds will be unusually attenuated (Unknown Carver Number 2); or possibly . . . well, we shall see.

If it is certain that Martha's Vineyard, an island whose furthest edges may be seen from its highest hill, produced these decoys and many, many more, it is also certain that equally numerous carvings were created in each of a long list of other coastal communities. If Monhegan Island, Maine, could give to the

144. Typical Vineyard decoy anchors.

world of art the stately and powerful sculpture of the scoters and eiders we are familiar with from earlier decoy literature, what of the whole long coast of Maine? What of Nova Scotia and the other Maritime Provinces? All of that northern coast is decoy territory and, unlike the areas further south where wildfowling has become the pursuit largely of men with time and money, it is still a seasonal necessity of life to the northern coastal fishing folk; a cheap source of food in a region of hard economic uncertainty. In Nova Scotia the sink-box is still in use with all of its attendant floating lures. Surely, among the pedestrian carving efforts, there must be here and there the work of men of imagination and talent, with new concepts of their lovely prey to astonish and delight our eyes.

The best bird decoys of many regions cannot easily fall into the all-inclusive category known as folk art. They are informed with the same searching honesty, the same knowledge of the sought-after game, as are the prehistoric paintings on the walls of the Spanish caves of Altimira and Lascaux, paintings which served as religious totems, as supplicant offers to the hunter gods to make the kill more certain.

I seem to hear the rather dusty laughter of the long-gone gunners, amused at the idea of having their decoy ducks described as Art. They were made only to toll the birds, not to leave for posterity the evidence of man's ability to create from common materials things of truth and occasional great beauty. The secret of the fascination of old decoys, of the endless variations of their small forms, lies somehow in the word unselfconscious. They pay no conscious homage, as do other art forms, to God or to man or to ego; their sole intent is to convince a living duck that they too are alive. Thus their rare integrity and, finally, their immortality.